World Future

Popular Review of new Consepts, Ideas and Innovations in Space Launch and Flight

(A popular presentation of the new scientific articles)

Dr.Sci., Professor **Alexander Bolonkin**,

The former senior scientific researcher of NASA and scientific laboratories of the Air Forces of the USA.

Lulu , USA, 2017

Title: **Popular Review of new Consepts, Ideas, and Innovations in Space launch and Flight**
Author: **Alexander Bolonkin**
ISBN 978-1- 365-98069-5

New macro-projects, concepts, ideas, methods, and innovations are explored here, but hardly developed. There remain many problems that must be researched, modeled, and tested before these summarized research ideas can be practically designed, built, and utilized—that is, fully developed and utilized.

Most ideas in our book are described in the following way: 1) Description of current state in a given field of endeavor. A brief explanation of the idea researched, including its advantages and short comings. No eny equations. Some of them have only the results of estimation and computations. But all ideas contains the links in initial scientific articles contains detail proofs, equtions and computations. 2) A brief description of possible applications—candidate macro-projects, including estimations of the main physical parameters of such economic developmental undertakings.

The parts are in a popular form accessible to the wider reading public. The many original articles of this book will require some mathematical and scientific knowledge, such as may be found amongst technical school graduate students.

The book gives the main physical data which will help researchers, engineers, dedicated students and enthusiastic readers make estimations for their own macro-projects. Also, inventors will find an extensive field of inventions and innovations revealed in our book.

The author have published many new ideas and articles and proposed macro-projects in recent years (see: References). This book is useful as an archive of material from the authors' own articles published during the last few years.

Every supparts is independent. Than why some figures are repited.

Copyright @ auther.
Publisher: USA LULU, www.lulu.com.

Contents

About Author
Abstract

Part A   7

1. Cable Space Launcher
2. Circle Launcher and Space Keeper
3. Kinetic Launcher and Kinetic keeper
4. Gas tube hypersonic launcher
5. Earth-Moon cable transport system
6. Earth-Mars cable transport system
7. Centrifugal space launchers
8. Asteroids as propulsion system of space ships
9. Multi-reflex propulsion systems
10. Electrostatic Solar wind propulsion
11. Electrostatic utilization of asteroids as space propulsion
12. Electrostatic levitation
13. Guided solar sail and energy generator
14. Radioisotope space sail and electro-generator
15. Electrostatic solar sail
16. Recombination space jet propulsion engine
17. Electronic sail

Part B   42

1. Electrostatic AB-ramjet space propulsion
2. Beam space propulsion
3. MagSail
4. High speed AB-solar sail
5. Transfer electricity in outer space
6. Simplest AB-thermonuclear space propulsion
7. Electrostatic linear engine and cable space launcher
8. AB-levitrons
9. Electrostatic climber
10. AB-space propulsion
11. Wireless transfer of energy
12. Magnetic space launcher
13. Railgun
14. Superconductivity rail gun
15. Convertor any matter in nuclear energy and photon rocket
16. Femtotechnology and its application into aerospace technology

Part C   111

1. Space Elevator, Transport System for Space Elevator
2. Men without the space suite into space
3. Electrostatic levitation and Artificial gravity
4. New method of atmospheric re-entry of space ship
5. Inflatable Dome for Moon, Mars, satellites, and space hotel
6. Ab method irrigation for planet without water. (Closed loop water cycle)
7. Artificial explosion of the Sun

About the Author

Alexander A. Bolonkin was born in the former USSR. He holds doctoral degree in aviation engineering from Moscow Aviation Institute and a post-doctoral degree in aerospace engineering from Leningrad Polytechnic University. He has held the positions of senior engineer in the Antonov Aircraft Design Company and Chairman of the Reliability Department in the Clushko Rocket Design Company. He has also lectured at the Moscow Aviation Universities. Following his arrival in the United States in 1988, he lectured at the New Jersey Institute of Technology and worked as a Senior Researcher at NASA and the US Air Force Research Laboratories.

Bolonkin is the author of more than 250 scientific articles and books and has 17 inventions to his credit. His most notable books include The Development of Soviet Rocket Engines (Delphic Ass., Inc., Washington , 1991); Non-Rocket Space Launch and Flight (Elsevier, 2006); New Concepts, Ideas, Innovation in Aerospace, Technology and Human Life (NOVA, 2007); Macro-Projects: Environment and Technology (NOVA, 2008); Human Immortality and Electronic Civilization, 3-rd Edition, Lulu, 2007; Life,Science,Future, Publish America, 2010): Small Non-Expensive Electric Cumulative Thermonuclear Reactors, Lulu, 2017.

Abstract

In recent years of the 21st Century the author of this book and other scientists as well, have instigated and described many new consepts, ideas, researches, theories, macro-projects, USA and other countries patented concepts, speculative macro-engineering ideas, projects and other general innovations in technology and environment change. These all hold the enticing promise for a true revolution in the lives of humans everywhere in the Solar System.

Here, the author includes and reviews new methods of space launch and flight. He include also topics having very omportant for space launch and flight. For example: converting of any matter into energy, getting of super strong materials, for travel in outer space without space suit, magnetic space launchers, magnetic space towers, motionless satellites and suspended structures, wireless transfer of electricity to long distance, Magnetic guns, magnetic launchers, space elevators and space climbers, along with many others.

Author succinctly summarizes in popular form the some of these revolutionary macro-projects, concepts, ideas, innovations, and methods for scientists, engineers, technical students, and the world public. Every subPart has two main supsections: At first subsection the author describes the new idea in an easily comprehensible way acceptable for the general public (no equations). One supsection also contains the scientific proof of the innovation acceptable for technical students, engineers and scientists, and the second supsection contains the applications of innovation.

Author does seek future attention from the general public, other macro-engineers, inventors, as well as scientists of all persuasions for these presented innovations. And, naturally, he fervently hopes the popular news media, various governments and the large international aerospace and other engineering-focused corporations will, as well, increase their respective observation, R&D activity in the technologies for living and the surrounding human environment.

Preface

New macro-projects, concepts, ideas, methods, and innovations are explored here, but hardly developed. There remain many problems that must be researched, modeled, and tested before these summarized research ideas can be practically designed, built, and utilized—that is, fully developed and utilized.

Most ideas in our book are described in the following way: 1) Description of current state in a given field of endeavor. A brief explanation of the idea researched, including its advantages and short comings; 2) Where methods, estimation and computations of the main system parameters are listed (initialussues), and 3) A brief description of possible applications—candidate macro-projects, including estimations of the main physical parameters of such economic developmental undertakings.

The first and third parts are in a popular form accessible to the wider reading public, the second part of this book will require some mathematical and scientific knowledge, such as may be found amongst technical school graduate students.

The book gives the main physical data and technical equations in attachments which will help researchers, engineers, dedicated students and enthusiastic readers make estimations for their own macro-projects. Also, inventors will find an extensive field of inventions and innovations revealed in our book.

The author have published many new ideas and articles and proposed macro-projects in recent years (see: General References). This book is useful as an archive of material from the authors' own articles published during the last few years.

Every supchapter is independent. Than why some figures are repited.

Acknowledgement

1. Some data in this work is garnered from Wikipedia under the Creative Commons License.
2. The author wish to acknowledge Joseph Friedlander, Richard Cathcart for help in editing of this book.

Popular Reviews of New Consepts, Ideas, Innovation in Space Launch and Flight
Part A

Abstract

In the past years the author and other scientists have published a series of new methods which promise to revolutionize the space propulsion systems, space launching and flight. These include the cable propulsion system, circle propulsion system and space keeper, kinetic propulsion system, gas-tube propulsion system, sliding rotary method, asteroid employment, electromagnetic accelerator, Sun and magnetic sail, solar wind sail, radioisotope sail, electrostatic space sail, laser beam propulsion system, kinetic anti-gravitator (repulsator), Earth-Moon non-rocket and Earth-Mars non-rocket transport system, multi-reflective beam propulsion system, electrostatic levitation, etc.

Some of them have the potential to decrease launch costs thousands of time, other allow to change the speed and direction of space apparatus without the spending of fuel.

The author reviews and summarizes some revolutionary propulsion systems for scientists, engineers, inventors, students and the public.

Key words: Review, Non-rocket propulsion, non-rocket space launching, non-rocket space flight, cable launch system, circle launch system, space keeper, kinetic propulsion system, gas-tube launch system, sliding rotary method, asteroid employment, electromagnetic accelerator, Sun and magnetic sail, solar wind sail, radioisotope sail, electrostatic space sail, laser beam propulsion system, kinetic anti-gravitator (repulsator), Earth-Moon non-rocket and Earth-Mars non-rocket transport system, multi-reflective beam propulsion system, electrostatic levitation, recombination engine, electronic sail, solar sail.

Introduction

Brief History. People have long dreamed to reach the sky. The idea of building a tower high above the Earth into the heavens is very old [1],[6]. The Greed Pyramid of Gaza in Egypt constructed c.2570 BCE has a height 146 m. The writings of Moses, about 1450 BC, in Genesis, Chapter 11, refer to an early civilization that in about 2100 BC tried to build a tower to heaven out of brick and tar. This construction was called the Tower of Babel, and was reported to be located in Babylon in ancient Mesopotamia. Later in chapter 28, about 1900 BC, Jacob had a dream about a staircase or ladder built to heaven. This construction was called Jacob's Ladder. More contemporary writings on the subject date back to K.E. Tsiolkovski in his manuscript "Speculation about Earth and Sky and on Vesta," published in 1895 [2-3]. Idea of Space Elevator was suggested and developed Russian scientist Yuri Artsutanov and was published in the Sunday supplement of newspaper "Komsomolskaya Pravda" in 1960 [4]. This idea inspired Sir Arthur Clarke to write his novel, The Fountains of Paradise, about a Space Elevator located on a fictionalized Sri Lanka, which brought the concept to the attention of the entire world [5].

Rockets for military and recreational uses date back to at least 13th century China.

Wernher von Braun, at the time a young aspiring rocket scientist, joined the military (followed by two former VfR members) and developed long-range weapons for use in World War II by Nazi Germany. In 1943, production of the V-2 rocket began in Germany. It had an operational range of 300 km (190 mi) and carried a 1,000 kg (2,200 lb) warhead, with an amatol explosive charge. It normally achieved an operational maximum altitude of around 90 km (56 mi), but could achieve 206 km (128 mi) if launched vertically.

After World War 2 the missile systems have received the great progress and achieved a great success. But rocket system is very expensive. In end of 1990 the researchers begin to study the non-rocket systems which promise to decrease the space launch and flight cost in hundreds times. The pioneer of these researches professor Alexander Bolonkin published the first serious book [1] in this field.

Current status of non-rocket space launch and flight systems. Over recent years interference-fit joining technology including the application of space methods has become important in the achievement of space propulsion system. Part results in the area of non-rocket space launch and flight methods have been patented recently or are patenting now.

Professor Bolonkin made a significant contribution to the study of the different types of non-rocket space launch and flight in recent years [1],[6]-[22] (1982-2011). Some of them are presented in given review.

Cable Space Launcher is researched in [1, pp.39-58]; Circle Launcher and Space Keeper were developed in [1, pp.59-82]; Kinetic Launcher and Kinetic Keeper were researched in [1, Ch.5, 9], [9], [10]; Gas tube hypersonic launcher presented in [1, pp.125-146]; Earth-Moon cable transport system was offered in [1, pp.147-156]; Earth-Mars cable transport system [1, pp.157-164]; Centrifugal space launchers were suggested in [1, pp.187-208, pp.223-244]; Asteroids as propulsion system of space ships were published in [1, pp.209-222]; Multi-reflex propulsion systems were researched in [1, pp.223-244]; Electrostatic Solar wind propulsion was developed in [1, pp.245-270]; Electrostatic utilization of asteroids as space propulsion in [1, pp.271-280]; Electrostatic levitation is presented in [1. pp.281-302]; Guided solar sail and energy generator is described in [1, pp.303-308]; Radioisotope space sail and electro-generator is presented in [1, pp.309-316]; Electrostatic solar sail is researched in [1, pp. 317-326]; Recombination space jet propulsion engine is described in [1, pp.327-340]; and Electronic sail is described in [1, pp.327-340]. Some of these systems were developed in [2]-[23].

Significant scientific, interplanetary and industrial use did not occur until the 20th century, when rocketry was the enabling technology of the Space Age, including setting foot on the moon.

But rockets are very expensive and have limited possibilities. In the beginning 21th century the researches of non-rocket launch and flight started [1], [5]-[8].Some of them are described in this review.

Main types of Non-Rocket Space Propulsion System

1. Cable Space Launcher
2. Circle Launcher and Space Keeper
3. Kinetic Launcher and Kinetic keeper
4. Gas tube hypersonic launcher
5. Earth-Moon cable transport system
6. Earth-Mars cable transport system

7. Centrifugal space launchers
8. Asteroids as propulsion system of space ships
9. Multi-reflex propulsion systems
10. Electrostatic Solar wind propulsion
11. Electrostatic utilization of asteroids as space propulsion
12. Electrostatic levitation
13. Guided solar sail and energy generator
14. Radioisotope space sail and electro-generator
15. Electrostatic solar sail
16. Recombination space jet propulsion engine
17. Electronic sail

1. Cable Space Launcher*

A method and facilities for delivering payload and people into outer space are presented. This method uses, in general, engines and a cable located on a planetary surface. The installation consists of a space apparatus, power drive stations located along the trajectory of the apparatus, the cable connected to the apparatus and to the power stations, and a system for suspending the cable. The drive stations accelerate the apparatus up to hypersonic speed.

The estimations and computations show the possibility of making these projects a reality in a short period of time (two project examples are given: a launcher for tourists and a launcher for payloads). The launch will be very cheap at a projected cost of $1–$5 per pound. The cable is made from cheap artificial fiber widely produced by modern industry.

----

*This chapter was presented (full scientific text) as Bolonkin's papers IAC-02-V.P.06, IAC-02-S.P.14 at World Space Congress-2002, Oct. 10–19, Houston, TX, USA, and as variant No. 8057 at symposium "The Next 100 years", 14–17 July 2003, Dayton, Ohio, USA; or see [1, pp.39-58].

Brief Description. The installation includes (see notations in Fig. 1.1): a cable; power drive stations; winged space apparatus (space ship, missile, probe, projectile and so on); conventional engines and flywheels; and a launching area (airdrome). Between drive stations the cable is supported by columns. The columns can also hold additional cables for future launches and a delivery system for used cable.

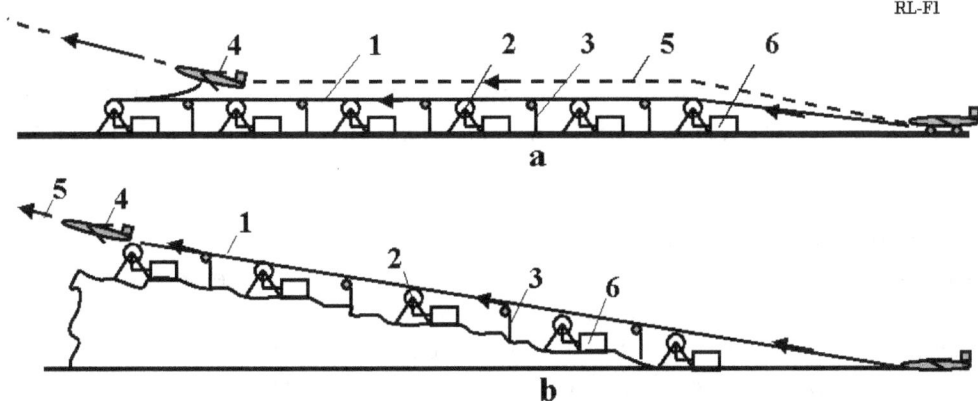

Fig. 1.1. *a*. Launcher for a crewed space ship with single cable. Notations: 1 – cable contains 3 parts: main part, outlet part, and directive part; 2 – power drive station; 3 – cable support columns; 4 – winged space

apparatus (space ship, missile, probe, projectile and so on); 5 – trajectory of space apparatus; 6 – engine. ***b***. A fixed slope small launcher for projectiles.

The installation works in the following way. All drive station start to run. The first power station pulls the cable, 1, connected to the winged space apparatus. The apparatus takes off from the start area and flies with acceleration along trajectory 5. When the apparatus reaches the first drive station, this drive station disconnects from the cable and the next drive station continues the apparatus acceleration, and so on. At the end of the distance, the winged apparatus has reached hypersonic speed, disconnects from the cable, changes the horizontal acceleration into vertical acceleration (while it is flying in the atmosphere) and leaves the Earth's atmosphere.

. The power stations contain the engines. The engine can be any type, for example, gas turbines, or electrical or mechanical motors. The power drive station has also an energy storage system (flywheel accumulator of energy), a type transmission and a clutch. The installation can also have a slope and launch a projectile at an angle to horizon (Fig. 1.1b).

2. Circle Launcher and Space Keeper*

The author proposes a new method and installation for flight in space. This method uses the centrifugal force of a rotating circular cable that provides a means to launch a load into outer space and to keep the stations fixed in space at altitudes at up to 200 km. The proposed installation may be used as a propulsion system for space ships and/or probes. This system uses the material of any space body for acceleration and changes to the space vehicle trajectory. The suggested system may also be used as a high capacity energy accumulator.

The article contains the theory of estimation and computation of suggested installations and four projects. Calculations include: a maximum speed given the tensile strength and specific density of a material, the maximum lift force of an installation, the specific lift force in planet's gravitation field, the admissible (safe) local load, the angle and local deformation of material in different cases, the accessible maximum altitudes of space cabins, the speed than a space ship can obtain from the installation, power of the installation, passenger elevator, etc. The projects utilize fibers, whiskers, and nanotubes produced by industry or in scientific laboratories.

----
* Detail manuscript was presented as Bolonkin's paper IAC-02-IAA.1.3.03 at the Would Space Congress-2002, 10-19 October, Houston, TX, USA. The material is published in *JBIS*, vol. 56, No 9/10, 2003, pp. 314-327. See in [1, pp.59-82].

Short description of Circle Launcher.
The installation includes (Fig. 2.1): a closed-loop cable made from light, strong material (such as artificial fibers, whiskers, filaments, nanotubes, composite material) and a main engine, which rotates the cable at a fast speed in a vertical plane. The centrifugal force makes the closed-loop cable a circle. The cable circle is supported by two pairs (or more) of guide cables, which connect at one end to the cable circle by a sliding connection and at the other end to the planet's surface. The installation has a transport (delivery) system comprising the closed-loop load cables (chains), two end rollers at the top and bottom that can have medium rollers, a load engine and a load. The top end of the transport system is connected to the cable circle by a

sliding connection; the lower end is connected to a load motor. The load is connected to the load cable by a sliding control connection.

The installation can have the additional cables to increase the stability of the main circle, and the transport system can have an additional cable in case the load cable is damaged.

The installation works in the following way. The main engine rotates the cable circle in the vertical plane at a sufficiently high speed so the centrifugal force becomes large enough to it lifts the cable and transport system. After this, the transport system lifts the space station into space.

The first modification of the installation is shown in Fig. 2.2. There are two main rollers 20, 21. These rollers change the direction of the cable by 90 degrees so that the cable travels along the diameter of the circle, thus creating the form of a semi-circle. It can also have two engines. The other parts are same.

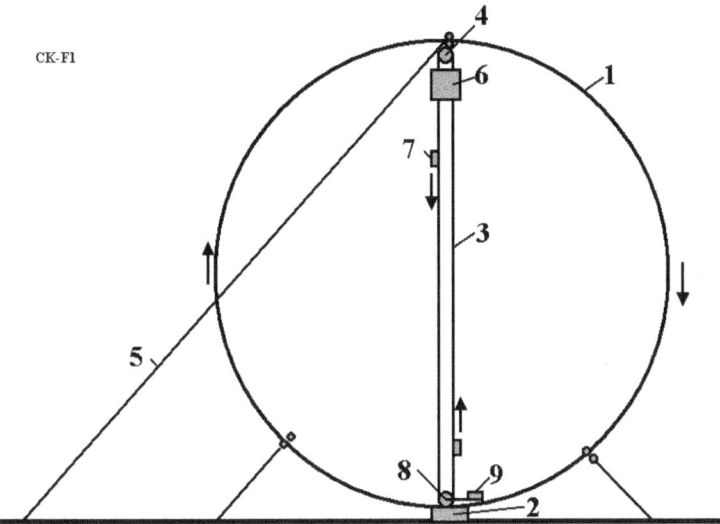

Fig. 2.1. Circle launcher (space station keeper) and space transport system. Notations are: 1 – cable circle, 2 – main engine, 3 – transport system, 4 – top roller, 5 – additional cable, 6 – the load (space station), 7 – mobile cabin, 8 – lower roller, 9 – engine of the transport system.

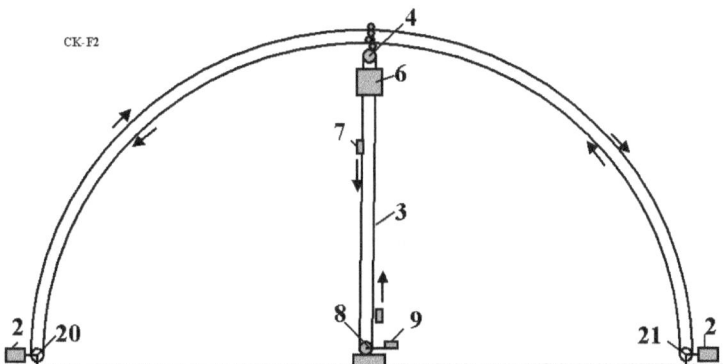

Fig. 2.2. Semi-circle launcher (space station keeper) and transport system. Notation is the same with Fig. 3.1 with the additional 20 and 21 – rollers. The semi-circle is the same (see right side of Fig. 3.4).

The installation can be used for the launch of a payload to outer space (Fig. 2.3). The load is connected to the cable circle by a sliding bearing through a brake. The load is accelerated by the cable circle, lifted to a high altitude, and disconnected at the top of the circle (semi-circle).

The installation may also be used as transport system for delivery of people and payloads from one place to another through space (Fig. 3.4 in [1]).

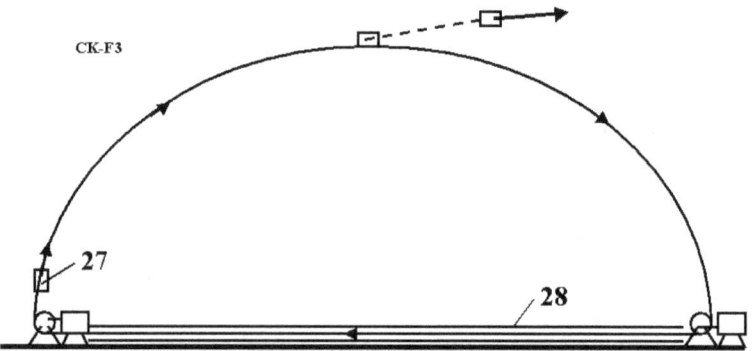

Fig. 2.3. Launching the space ship (probe) into space using cable semi-circle. 27 – load, 28 – vacuum tube (option).

This system works in the following way. The installation has two cable circles, which move in the opposite directions at the same speed. The space stations are connected to the cable circle through the sliding connection. They can move along the circle in any direction when they are connected to one of the cable circles through a friction clutch, transmission, gearbox, brake, and engine, and can use the transport system in Figs. 2.1 and 2.2 for climbing to or descending from the station. Because energy can be lost through friction in the connections, the energy transport system and drive rollers transfer energy to the cable circle from the planet surface. The cable circles are supported at a given position by the guide cables (see Project 2 in [1, Ch.3)]. No towers for supporting the circle cable are needed.

The system can have only one cable (Figs. 2.1, 2.3).

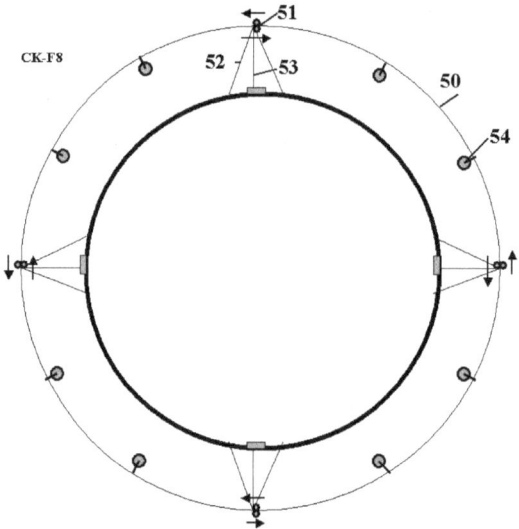

Fig. 2.4. Cable circle around the Earth for 8–10 space objects. Notations are: 50 – double circle, 51 – drive stations, 52 – guide cable, 53 – energy transport system, 54 – space station.

The installation can have a system for changing the radius of the cable circle ([1], Fig. 3.9). When an operator moves the tackle block, the length of the cable circle is changed and the radius of the circle is also changed.

3. Kinetic launcher on kinetic towers*

The author discusses a revolutionary new method to access outer space. A cable stands up vertically and pulls up its payload into space with a maximum force determined by its strength. From the ground the cable is allowed to rise up to the required altitude. After this, one can climb to an altitude using this cable or deliver a payload at altitude. The author shows how this is possible without infringing the law of gravity.

The original article contains the theory of the method and the computations for four projects for towers that are 4, 75, 225 and 160,000 km in height. The first three projects use the conventional artificial fiber widely produced by current industry, while the fourth project use nanotubes made in scientific laboratories. The chapter also shows in a fifth project how this idea can be used to launch a load at high altitude.

----

*Presented as paper IAC-02-IAA.1.3.03 at Would Space Congress 2002, 10–19 October, Houston, TX, USA. Detail manuscript was published as Bolonkin, A.A. "Kinetic Space Towers and Launchers", *JBIS*, Vol. 57, No.1/2, 2004, pp.33-39. Or see in [1, Ch.5, pp.107-124].

Brief description of innovation.

The installation (kinetic tower) includes (see notations in Fig. 3-1a,b and others): a strong closed-loop cable, two rollers, any conventional engine, a space station, a load elevator, and support stabilization ropes.

The installation works in the following way. The engine rotates the bottom roller and permanently sends up the closed-loop cable at high speed. The cable reaches a top roller at high altitude, turns back and moves to the bottom roller. When cable turns back it creates a reflected (centrifugal) force. This force can easily be calculated using centrifugal theory, or as reflected mass using a reflection theory. The force keeps the space station suspended at the top roller; and the cable (or special elevator) allows the delivery of a load to the space station. The station has a parachute that saves people if the cable or engine fails.

The theory shows, that current widely produced artificial fibers (see References[1] for cable properties) allow the cable to reach altitudes up to 100 km (see Projects 1 and 2 in [1] Ch.5). If more altitude is required a multi-stage tower must be used ([1], Fig. 5.2, see also Project 3 in [1] Ch.5). If a very high altitude is needed (geosynchronous orbit or more), a very strong cable made from nanotubes must be used (see Project 4).

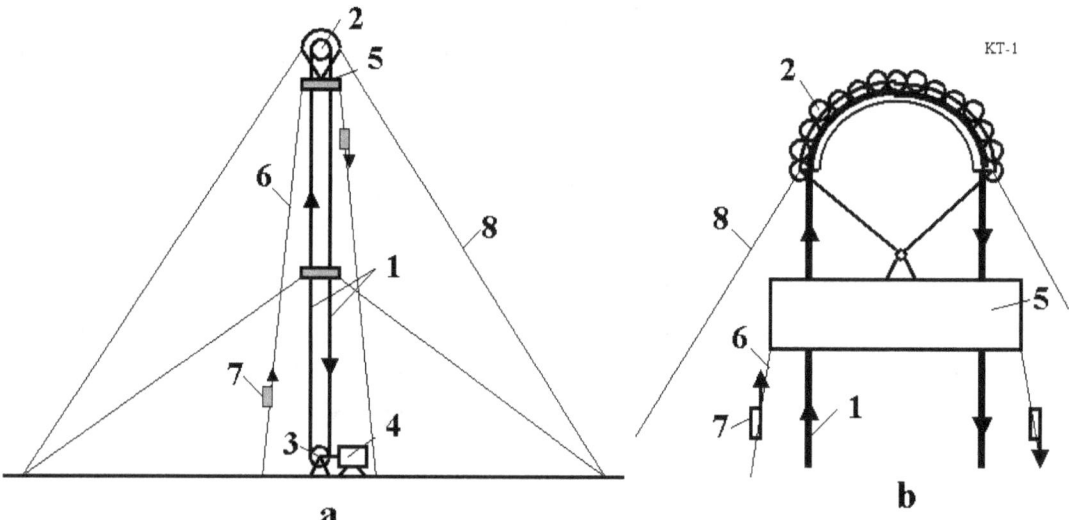

Fig. 3.1. **a.** Offered kinetic tower: 1 – mobile closed loop cable, 2 – top roller of the tower, 3 – bottom roller of the tower, 4 – engine, 5 – space station, 6 – elevator, 7 – load cabin, 8 – tensile element (stabilizing rope). **b.** Design of top roller.

The offered tower may be used for a horizon launch of the space apparatus (Fig. 3.2). The vertical kinetic towers support horizontal closed-loop cables rotated by the vertical cables. The space apparatus is lifted by the vertical cable, connected to horizontal cable and accelerated to the required velocity.

The closed-loop cable can have variable length. This allows the system to start from zero altitude, and gives the ability to increase the station altitude to a required value, and to spool the cable for repair. The device for this action is shown in [1], p.110, Fig. 5.4. The offered spool can reel in the left and right branches of the cable at different speeds and can change the length of the cable.

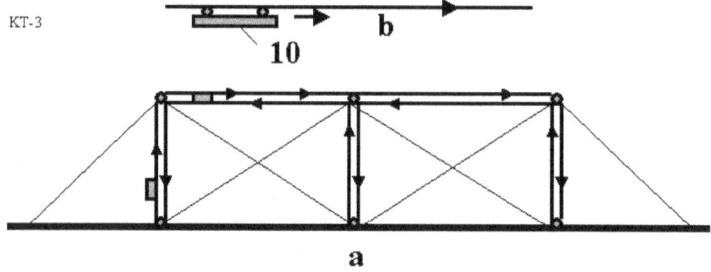

Fig. 3.2. **a.** Kinetic space installation with horizontally accelerated parts. **b.** 10 – accelerated missile.

4. Gas Tube Hypersonic Launcher*

The present review describes a hypersonic gas rocket, which uses tube walls as a moving compressed air container. Suggested burn programs (fuel injection) enable use of the internal tube components as a rocket. A long tube (up to 0.4–0.8 km) provides mobility and can be aimed in water. Relatively inexpensive oxidizer and fuel are used (compressed air or gaseous oxygen and kerosene). When a projectile crosses the Earth's atmosphere at an angle more than 15°, loss of speed and the weight of the required thermal protection system are small. The research shows that the launcher can give a projectile a speed of up to 5–8 km/s. The proposed launcher can deliver up to 85,000 tons of payloads to space annually at a cost of one to two dollars per pound of payload. The launcher can also deliver about 500 tons of mail or express parcels per day over continental distances and may be used as an energy station and accumulator. During war, this launch system could deliver military munitions to targets thousands to tens of thousands of kilometers away from the launch site.

----

* This review is based on a paper presented at the 38th AIAA Propulsion Conference, 7–10 July 2002, Indianapolis, USA (AIAA-2002-3927) and the World Space Congress, 10–19 Oct. 2002, Houston, USA (IAC-02-S.P.15). Detailed material is published as A.A.Bolonkin, "Hypersonic Gas-Rocket Launcher of High Capacity", *JBIS*, vol. 57, No. 5/6, 2004, pp. 162–172; Journal *Actual Problems of Aviation and Aerospace Systems*, Kazan, 1 (15), pp. 45-69, 2003.
See also in [1, Ch.6, pp.125-146].

Description

Fig. 4.1a shows a design of the tube of the suggested hypersonic gas-rocket system. The system is made up of a tube, a piston with a fuel tank and payload, and nozzle connected to the piston, and valves.

The tube rocket engine can be made without a special nozzle (Fig. 4.1b). In this case, the fuel efficiency of the gas-rocket engine will decrease but its construction becomes simpler.

The tube may be placed into a frame (Fig. 4.1c). The frame is placed into water and connected to a ship for mobility and aiming.

The launch sequence is as follows. First the movable piston with the fuel tank (containing liquid fuel), and payload are loaded into the tube. The piston is held in place by the fasteners or closed valve 17 (Fig. 4.1). The direction and angle of the launch tube are set.

Valve 19 (Fig. 4.1) is closed and a vacuum (about 0.005 atm) is created in the launch tube space above the payload/piston to reduce the drag imparted to the payload/piston as it moves along the launch tube. The tube, of a length of 630 m and a diameter of 10 m, contains 61 tons of air at atmospheric pressure. If this air is not removed, the payload must be decreased by the same value.   If air pressure is decreased down to 0.005 atm, the parasitic air mass is decreased to 300 kg. This is an acceptable parasitic load.

Valve 17 is closed and an oxidizer (air, oxygen, or a mixture) is pumped into the space below the payload/piston.

Liquid fuel (benzene, kerosene) is injected into the space below nozzle 6 through the launch tube injectors (item 15, Fig. 4-1) and ignited. Valve 17 (Fig. 4-1) is opened. The hot combustion gas expands and pushes the payload/piston system along the launch tube together with the air column (item 7) between the piston and nozzle.

When the piston reaches the maximum gun speed (about 1 km/s), the compressed air column begins to work as a rocket engine using one of the special injection fuel programs (see Reference in [1, Ch.6[12]]).

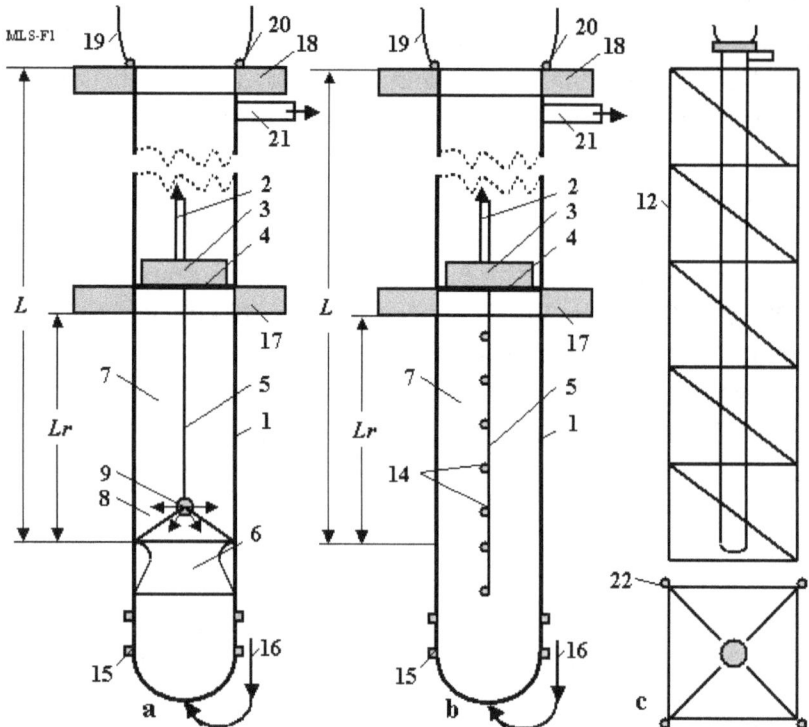

Fig. 4.1. a - Space launcher with the gas rocket and rocket nozzle in the tube. The system comprises the following: 1 – tube, 2 – payload (projectile), 3 – fuel tank, 4 – piston, 5 – fuel pipeline, 6 – nozzle connected to piston, 7 – rocket air column, 8 – combustion chamber, 9 – injectors of the combustion chamber, 12 – tube frame, 14 – additional injectors, 15 – lower tube injectors, 16 – air pipeline, 17 – lower valve, 18 – upper valve, 19 – top valve, 20 – air lock, 21 – gas pipe, 22 – electric engines.
b - Space launcher with the gas rocket and no the rocket nozzle. c – Launcher in frame.

As the payload/piston approaches the end of the launch tube, valve 19 is opened and the airlock (item 20) begins to operate. After the payload/piston has left the launch tube, valve 18 closes the end of the launch tube and re-directs the hot combustion gases down the bypass tube (item 21) to various turbo-machines preparing compressed air for the next shot and electricity for customers.

If a high launch frequency is required, then internal tube water injectors are used to quickly cool the launch tube.

After the payload/piston system leaves the launch tube, the payload (projectile) separates from the piston and the empty fuel tank. The payload continues to fly along a ballistic trajectory. At apogee, the payload may use a small rocket engine to reach orbit or to fly to any point on Earth.

The method by which the fuel is injected and ignited within the launch tube is critical to high-speed (hypersonic) acceleration of the payload. The author has developed the five fuel injection programs for the launch system[12].

In these programs the thrust (force) is constant at all times, which means that pressure and all parameters in the rocket engine are constant. Parts of the programs have two steps. In the first step the fuel is injected into compressed air at the lower part of the tube to support a constant pressure and provide the initial acceleration of the rocket (together with air column L_r)

to the velocity V_o. In the second step the rocket engine begins to thrust and support the constant pressure and temperature in the rocket combustion chamber. The result is that the thrust force of the gas-rocket engine remains constant. In the reference article the author considered only a simplified model ([1, Fig. 6.1b]) when a rocket nozzle is absent.

5. Earth–Moon Cable Transport System*

The author proposes a new transportation system for travel between Earth and the Moon. This transportation system uses mechanical energy transfer and requires only minimal energy, using an engine located on Earth. A cable directly connects a pole of the Earth through a drive station to the lunar surface. The equation for an optimal equal stress cable for the complex gravitational field of the Earth–Moon has been derived that allows significantly lower cable masses. The required strength could be provided by cables constructed of carbon nanotubes or carbon whiskers. Some of the constraints on such a system are discussed.

----

* This review is based on paper B0.3-F3.3-0032-02 that was presented to 34th COSPAR Scientific Assembly, The World Space Congress 2002, 10–19 Oct 2002, Houston, Texas, USA. This is only part of the original manuscript (one version of the system) presented to the WSC. This part of WSC manuscript was published in as "Non-Rocket Earth–Moon Transport System", in *Advanced Space Research*, Vol. 31, No. 11, pp. 2485–2490, 2003. See also [1, Ch.7, pp.147-155].

Brief Description.

The objectives of the proposed system are to provide an inexpensive means of travel between the Earth and the Moon, to simplify space transportation technology, and to eliminate complex hardware. The proposed Earth–Moon cable transport system is shown in Fig. 5.1. The system consists of three cables: a main (central) cable, which supports the weight of the entire system, and two closed-loop transport cables, which include a set (5–10) of cable chain links connected sequentially to one other by rollers [1, Ch.7[3, 4] (see Fig.7.3a)]. The system is connected at the Earth's pole and to any position on the Moon's surface that continually faces Earth. An engine located on a planet (e.g. the Earth, but it could be the Moon) drives the cable transport system. On the Earth, the cable is supported in the atmosphere by a winged device, which also counteracts the rotation of the Earth. The transfer cable system transfers energy between load cabins moved up and down, which requires the engine moving the cable system to overcome only frictional forces.

An optimal (minimum mass, equal stress) variable diameter cable is defined for the main tether. The main cable has a relatively large but variable cross-section area (diameter) because it has to support the total system weight, which is several hundred times the load weight. In an optimal main cable the cross-section area increases (for K = 2, about 20 times) in the altitude range 0 – 150,000 km, is approximately constant in the range 150,000–380,000 km, and decreases near the Moon's surface, from 20,000 km to the surface.

The mass of the main cable is minimized because its diameter is variable along the distance (see the next section for calculation of the main cable cross-section areas and mass). The transport cables pull (move) the load cabins (one up, the other down) along the main cable. As these are moveable parts, they must have constant diameter. If they had to carry a load the full distance to the Moon, their mass would be very large. My concept separates the full distance into sub-distances (5–10), with closed-loop links for every sub-distance connected by rollers. These rollers transfer the transport cable movement from one link to another. In this case, the mass of the transport cables is minimized because at every local length (sub-distance) the cable diameter is determined by the local force. Total mass of the transport cable should be close to double the mass of the main cable.

The load containers are connected to the transport cable. When containers come up to the rollers, they pass the rollers, connect to the next link and continue their motion along the main cable. The load (cabin) has special clamps to allow this transfer between the different diameter cables in each link[1]. Most space payloads, like tourists, must be returned to Earth. When one container is moved up, another container is moved down. The work of lifting equals the work of descent, except for a small friction loss in the rollers. The transport system may be driven by a conventional motor located at the Earth drive station, on a space station, or on the Moon. When payloads are not being delivered into space, the system may be used to transfer mechanical energy to the Moon. For example, the Earth drive station can rotate an electric generator on the Moon.

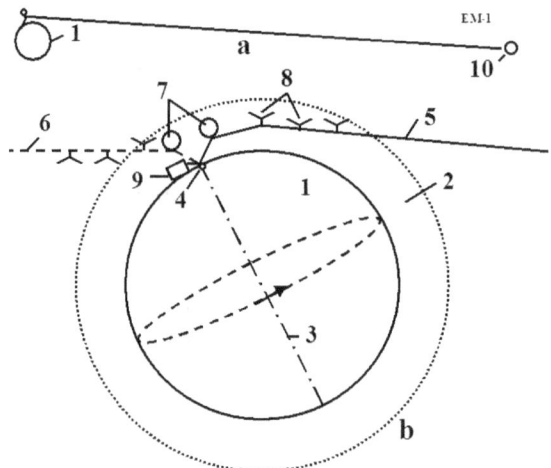

Fig. 5.1. A conceptual Earth–Moon transportation system. One end is connected to the Earth's pole. the second end is connected to the Moon. Notation: 1 – the Earth; 2 – Earth's atmosphere; 3 – axis of Earth rotation; 4 – Earth Pole; 5 – Earth–Moon cable transport system in right position (one extreme of the Moon's position); 6 – Earth–Moon system in left position; 7 – air balloons; 8 – support wings; 9 – drive station; 10 – Moon.

The cable is supported in the Earth's atmosphere by air balloons (around the pole) and winged devices (far from the pole). The maximum speed of the system in the atmosphere is about 190 m/sec at the maximum distance of 2700 km in the right-hand position of Fig. 5-1. When the cable is located in the left-hand position, some wings may be out of the atmosphere and not so effective.

The Moon's orbit has eccentricity. Every 29 days the Moon's distance from Earth changes by about 50,000 km. Devices shown in Fig. 7.4 (in [1, Chapter 7]) must be used to change the length (or link length) of the transport cables as the Earth–Moon distance changes. They may be located at the Earth drive station, on a space station, in space, and/or on the Moon. The average speed of a cable length change is about 40 m/s. As the Moon pulls the transport system, it may be used to produce mechanical energy. If the cables can support 9 tons, the power can reach 1.8 million Watts. The cables rotate the electric generator and negligibly brake the Moon's movement.

1. Earth–Mars Cable Transport System*

The author offers and computes a new permanent cable transport system that links a pole of the Earth with Mars orbit. This system connects Earth and Mars for 1–1.5 months every 1.7–2 years when they are located at the nearest distance and allows the transfer of people and loads to Mars and back. The system has many advantages because it uses a transport engine located on Earth, but it also requires the high strength cable made from nanotubes. This work contains theory of an optimal equal stress cable, that connects the Earth and Mars orbit, as well as computed parameters of the suggested system.

---

*Presented as paper BO.4-C3.4-0036-02 to The World Space Congress-2002 10–19 Oct. 2002, Houston, Texas, USA. Detailed material was published in *Actual Problems of Aviation and Aerospace Systems*. No. 2 (16), vol. 8, 2003. See also [1, Ch.8, pp.157-164].

Brief Description

The review contains the theory and results of computation for a special project. This project uses three cables (one main cable and two for driving loads) mass from artificial material: whiskers, nanotubes, with the specific tensile strength (ratio of tensile stress to density) $k = \sigma/\gamma = 20 \cdot 10^7$ ($K = 20$) or more. Nanotubes with the same or better parameters are available in scientific laboratories. The theoretical limit of nanotubes of SWNT type is about $k = 100 \cdot 10^7$ ($K = 100$).

A proposed centrifugal Mars cable transport system is shown in Figs. 6.1 and 6.2. The system includes the optimal equal stress cable which has a length approximately equal to the minimum distance of the Earth to Mars orbit. The installation has a transport system with chains connected by rollers and two transport cables.

The upper ends of the cables are located near Mars orbit and the lower ends of the cables are connected to the Earth's pole. They are supported in the Earth's atmosphere by air balloons (near the Pole) and winged devices at a maximum distance of up to 2800 km. The rotary speed of the cables changes from zero (at the pole) to 190 m/s (at the end of the maximum distance in the atmosphere). These winged devices can support cables when they are located within the lower atmosphere.

The installation would have a device that allows the length of the cables to be changed. The device would consist of a spool, motor, brake, transmission, and controller. The facility could have mechanisms for delivering people and payloads to Mars and back using the suggested transport system.

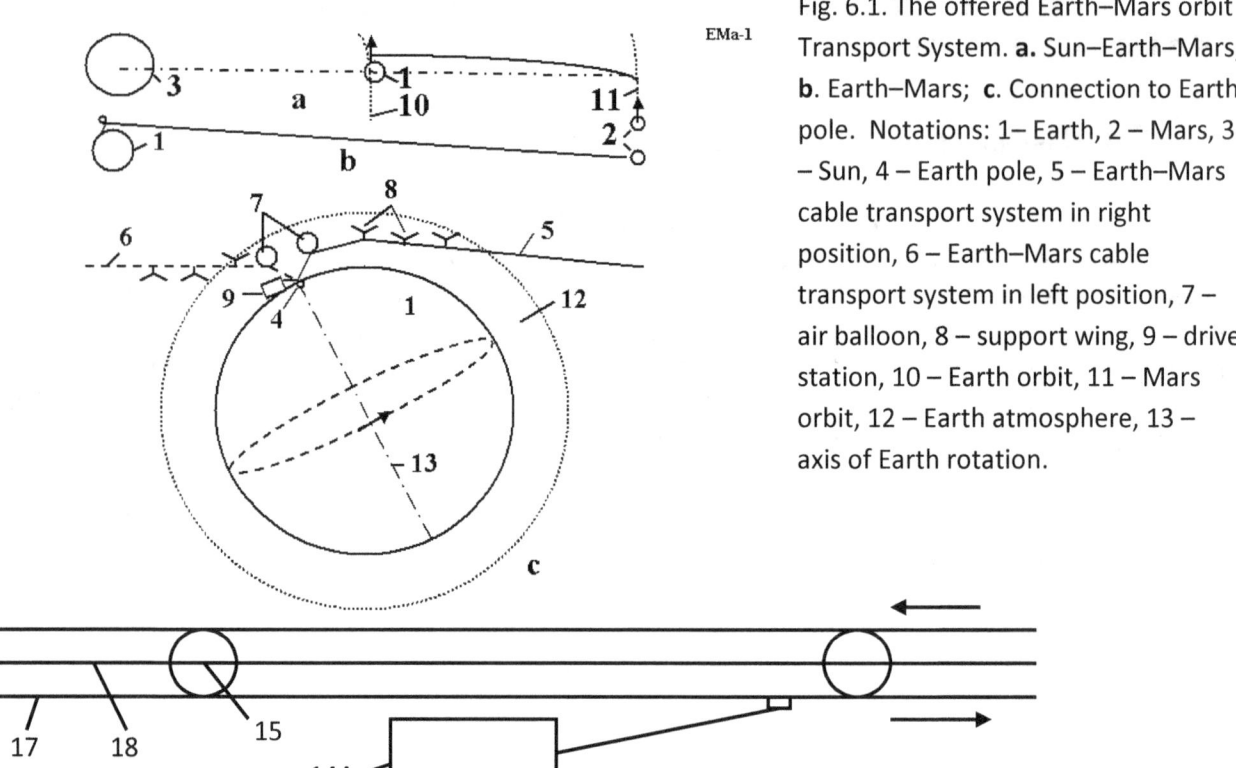

Fig. 6.1. The offered Earth–Mars orbit Transport System. **a.** Sun–Earth–Mars; **b.** Earth–Mars; **c.** Connection to Earth pole. Notations: 1– Earth, 2 – Mars, 3 – Sun, 4 – Earth pole, 5 – Earth–Mars cable transport system in right position, 6 – Earth–Mars cable transport system in left position, 7 – air balloon, 8 – support wing, 9 – drive station, 10 – Earth orbit, 11 – Mars orbit, 12 – Earth atmosphere, 13 – axis of Earth rotation.

Fig. 6.2. Cables of transport system. Notations: 144 – space ship, 15 – rollers, 17 – transport cable, 18 – main cable.

The delivery devices include: containers (passenger cabins, space ships, etc.), cables, motors, brakes, and controllers.

The space cabin can temporarily land on the surface of the Mars for loading and performing research. The space cabin has a small rocket engine for maneuvering and landing on the surface of Mars.

Every two years Mars comes within a minimum distance from Earth. For about 1–1.5 months the cable transport system (CTS) can be used to deliver people and loads to Mars. The space ship moves in advance to the upper end of the CTS, then when Mars arrives, the vehicles land on its surface and the people work on for Mars 1–1.5 months; afterwards the space ships return to Earth. While living on Mars, the people can fly from one place to another with speeds of about 230 m/s (including Mars round trip at low altitude) in their space cabin (ship). Exploring using the CTS would not require rocket fuel.

7. Centrifugal Space Launcher*

This manuscript describes a method and devices that provides a repulsive (repel, push, opposed to gravitation) force between given bodies. The basic concept is that a strong, heavy cable is projected upwards using a motorized wheel on the ground. The upward momentum of the cable is transferred to the apparatus by means of a pulley/roller mechanism, which sends the cable back down to the motor. The momentum transferred from the cable to the apparatus produces a push force which can suspend the apparatus in the air or lift it. There is an equal and opposite force on the motorized wheel on the ground. The push force can be great (up to tens of tons) and operate over long distances (up to hundreds of kilometers). This force produces great accelerations and velocities of given bodies (vehicles).

----

*The main idea of this Chapter was presented as IAC-02-IAA.1.3.03, 53rd International Astronautical Congress. The World Space Congress – 2002, 10–19 Oct. 2002, Houston, Texas, USA, and the full manuscript accepted as AIAA-2005-4504, 41 Propulsion Conference, 10–12 July, 2005, Tucson, Arizona, USA. See also [1, Ch.10, pp.187-206].

Description of Innovative Launcher

Ground sling launcher. The installation includes (see notations in Fig. 7.1): a tower, a lever (or disk), a sling (cable), conventional engines and flywheels (drive station). Optimally, the installation is located on a mountain (high altitude) to reduce air drag on the sling and apparatus and for a lower slope of initial trajectory angle. A winged space apparatus (space ship, missile, probe, projectile, etc.) is connected to the end of the sling.

The installation works in the following way (Fig. 7.1). The engine rotates the flywheels. When the flywheels accumulate sufficient energy, they rotate as a lever. The lever accelerates the space apparatus ("s.a."). The apparatus may be located on the lever and the sling is increased after the start.   The apparatus speed increases. It is greater than the lever speed in the ratio R/R_0, where R is the radius of the apparatus trajectory circle, R_0 is the radius of the lever (or disk). When the apparatus reaches its chosen speed, the winged apparatus is disconnected from the sling at the desired point of the circle. While the winged apparatus is flying in the atmosphere, it can increase its slope and correct its trajectory. If the apparatus has a hypersonic (supersonic) form, the speed loss is small[13].

The offered launcher is different from conventional centrifugal catapults, which have a projectile in a lever. This launcher has a long sling and the projectile is in the sling. The sling increases the lever speed many times and decreases the mass of the lever. Conventional catapults made from nanotubes have a huge mass and requires gigantic energy to work. This sling is also made from nanotubes (for space speed), but its mass is small.

If the circle is parallel to the Earth's surface, the winged apparatus disconnects from the cable, converts the linear and centrifugal acceleration into vertical acceleration (while it is flying in the atmosphere) and leaves the Earth's atmosphere.

The power station houses the engine. It can be any engine, for example, a gas turbine, or an electrical or mechanical motor. The power drive station also has an energy storage system (flywheel accumulator of energy), a transmission drive train and a clutch mechanism.

The installation can be set on a slope, and launch a projectile at an angle to the horizon (Fig. 7.1).

The attained speed may be up to eight or more km/s (see project 2 below). If the planet does not have an atmosphere, a small installation (with a small lever) can give the projectile a very high speed, limited only by the power of the engine and the strength of the sling.

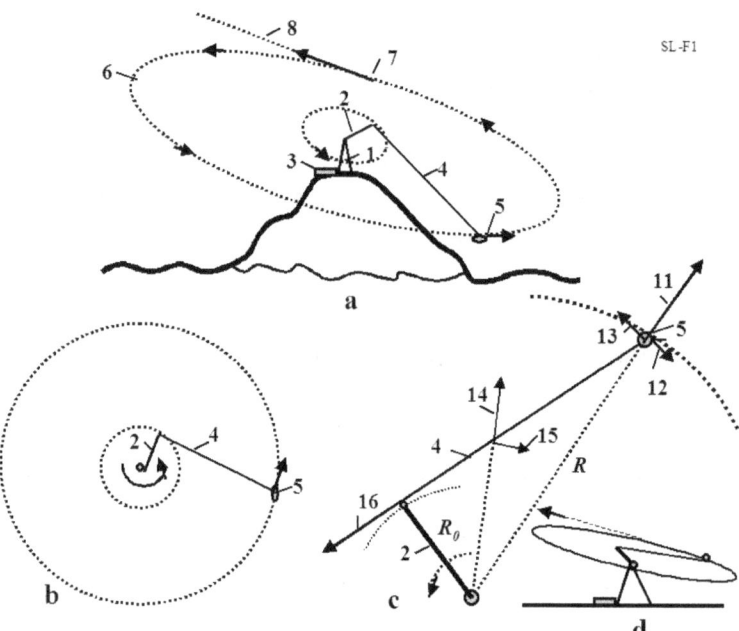

Fig. 7.1. Sling rotary launcher. a) launcher located on mountain, b) top view of installation, c) acting forces, d) side view. Notations: 1 – tower, 2 – lever or disk, 3 – engine, 4 – sling, 5 – space apparatus (s.a.), 6 – circular launch trajectory, 7 – point of disconnection, 8 – direction of launch, 11 – centrifugal force of space apparatus, 12 – drag of s.a., 13 – speed of s.a., 14 – centrifugal force of sling, 15 – drag of sling, 16 – lever force.

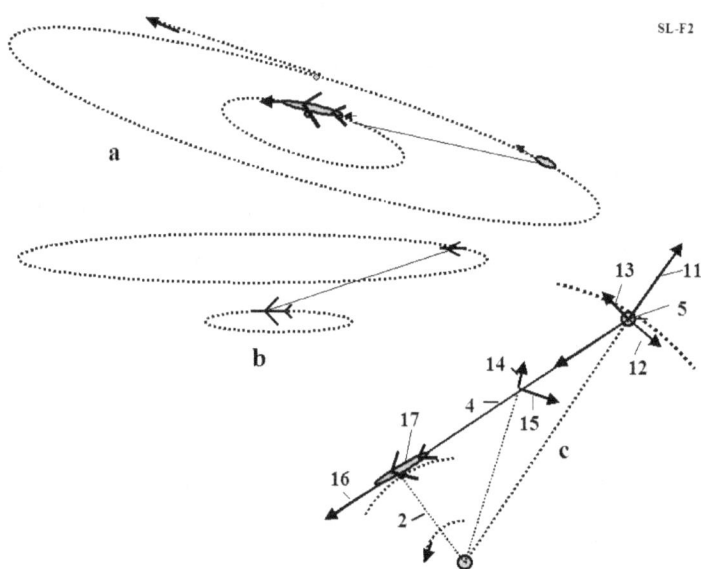

Fig. 7.2. Launching a space ship using aircraft. a) slinging slope start, b) upper start, c) installation forces.

On the Earth's surface the launcher can be located under a special cover (or in a tube) in a vacuum.

Aircraft sling launcher. Another design of this sling launcher is presented in Figs. 7.2. A small spacecraft (1 – 2 tons) is connected to a large, high-speed aircraft. The aircraft flies in a circle, increasing the sling length and accelerating the ship to high speed. The attained speed depends largely on the specific strength of the sling, the maximum aircraft speed and the thrust of the aircraft. For large existing aircraft operating in the atmosphere, the launch speed may reach up 2 km/s. This is enough for the X-prize flight, reaching an altitude of up to 100 km and sufficient for a spaceship for tourists (see projects 3–4 below).

Advantages. The suggested launch cable system has advantages compared to the current rocket systems, as follows:
1. The sling launcher is many times less expensive than modern rocket launch systems. No expensive rockets are needed. Only motor and cable are required.
2. The sling launcher reduces the delivery cost by several thousand times (to as low as $5 to $10 per pound). (No rocket, cheaper fuel.)
3. The sling launcher could be constructed within one to two years. The aircraft sling launcher requires only a cable and a spaceship. Modern rocket launch systems require many years for R&D and construction.
4. The sling launcher does not require high technology and can be made by any non-industrial country.
5. Rocket fuel is expensive. The ground sling launcher can use the cheapest sources of energy, such as wind, water, or nuclear power, or the cheapest fuels such as gas, coal, peat, etc., because the engine is located on the Earth's surface. Flywheels may be used as an accumulator of energy.
6. It is not necessary to have highly qualified personnel, such as rocket specialists with high salaries.
7. The fare for space tourists would be small.
8. There is no pollution of the atmosphere from toxic rocket gas.
9. Thousands of tons of useful payloads can be launched annually.

Shortcomings of sling space launchers:
1. The need for a very strong sling (cable), made from carbon whiskers or still-to-be manufactured long nanotubes.
2. The Earth ground sling launcher may be used only for robust loads because high centrifugal acceleration is imposed on the payload. Such payloads normally account for 70–80% of space payloads.

Cable (sling) discussion. The experimental and industrial fibers, whiskers, and nanotubes are considered in [1], Chapters 1–2.

The reader can find a more complete cable discussion of cable and cable characteristics in [1], Ch.10, the References[3–13, 17–20].

8. Asteroids as Propulsion System of Space Ships*

The purpose of this section is to draw attention to the idea of sling rotary launchers. This idea allows the building of inexpensive new space launcher systems, to launch missiles, projectiles, and space apparatus, and to use many types of energy. This chapter describes the possibilities of this method and the conditions which influence its efficiency. Included are four projects: a non-rocket sling projectile launcher, a space sling launcher, a spaceship for launching using conventional supersonic, and a space ship using subsonic aircraft. The last two only require low-cost cable made from artificial fiber, using whiskers that are produced in industry now or increasingly perfected nanotubes that are being created in a scientific laboratories.

----

*The detailed work was presented as AIAA-2005-4035 at the 41 Propulsion Conference, 10–12 July, 2005, Tucson, Arizona, USA. See also [1, Ch.11, pp. 209-222].

Introduction.

There are many small solid objects in the Solar System called asteroids. The vast majority are found in a swarm called the asteroid belt, located between the orbits of Mars and Jupiter at an average distance of 2.1 to 3.3 astronomical units (AU) from the Sun. Scientists know of approximately 6,000 large asteroids of a diameter of 1 kilometer or more, and of millions of small asteroids with a diameter of 3 meters or more. Ceres, Pallas, and Vesta are the three largest asteroids, with diameters of 785, 110 and 450 km (621, 378, and 336 miles), respectively. Others range all the way down to meteorite size. In 1991 the Galileo probe provided the first close-up view of the asteroid Caspra; although the Martian moons (already seen close up) may also be asteroids, captured by Mars. There are many small asteroids, meteorites, and comets outside the asteroid belt. For example, scientists know of 1,000 asteroids of diameter larger than one kilometer located near the Earth. Every day 1 ton meteorites with mass of over 8 kg fall on the Earth. The orbits of big asteroids are well known. The small asteroids (from 1 kg) may be also located and their trajectory can be determined by radio and optical devices at a distance of hundreds of kilometers.

Radar observations enable to discern of asteroids by measuring the distribution of echo power in time delay (range) and Doppler frequency. They allow a determination of the asteroid trajectory and spin and the creation of an asteroid image.

Most planets, such as Mars, Jupiter, Saturn, Uranus, and Neptune have many small moons that can be used for the proposed space transportation method.

There are also the asteroids located at the stable Lagrange points of the Earth–Moon system. These bodies orbit with the same speed as Jupiter, and might be very useful for propelling spacecraft further out into the solar system. Comets may also be useful for propulsion once a substantial spacecraft speed is obtained. It seems likely that the kinetic and rotational energy of both comets and asteroids will eventually find application in space flight.

Most asteroids consist of carbon-rich minerals, while most meteorites are composed of stony-iron.

The present idea [1][6–8] is to utilize the kinetic energy of asteroids, comets, meteorites, and space debris to change the trajectory and speed of space ships (probes). Any space bodies more than 10% of a ship's mass may be used, but here mainly bodies with a diameter of 2 meters (6 feet) or larger are considered. In this case the mass (20–100 tons) of the space body (asteroid) is some 10 times more than the mass of probe (1 ton, 2000 lb) and the probe mass can be disregarded.

Connection Method

The method includes the following main steps:

(a) Finding an asteroid using a locator or telescope (or looking in catalog) and determining its main parameters (location, mass, speed, direction, rotation); selecting the appropriate asteroid; computing the required position of the ship with respect to the asteroid.
(b) Correcting the ship's trajectory to obtain the required position; convergence of the ship with the asteroid.
(c) Connecting the space apparatus (ship, station, and probe) to the space body (planet, asteroid, moon, satellite, meteorite, etc.) by a net, anchor, and a light strong rope (cable), when the ship is at the minimum distance from the asteroid.
(d) Obtaining the necessary position for the apparatus by moving around the space body and changing the length of the connection rope.
(e) Disconnecting the space apparatus from the space body; spooling the cable.

The equipment required to change a probe (spacecraft) trajectory includes:
(a) A light strong cable (rope).
(b) A device to measure the trajectory of the spacecraft with relative to the space body.
(c) A device for spacecraft guidance and control.
(d) A device for the connection, delivery, control, and disconnection and spooling of the rope.

Description of Utilization

The following describes the general facilities and process for a natural space body (asteroid, comet, meteorite, or small planet) with a small gravitational force to change the trajectory and speed of a space apparatus.

Figs. 8.1a,b,c show the preparations for using a natural body to change the trajectory of the space apparatus; for example, the natural space body 2, which is moving in the same direction as the apparatus (perpendicular to the sketch, Fig. 8.1a). The ship wants to make a maneuver (change direction or speed) in plane 3 (perpendicular to the sketch), and the position of the apparatus is corrected and moved into plane 3. It is assumed that the space body has more mass than the apparatus.

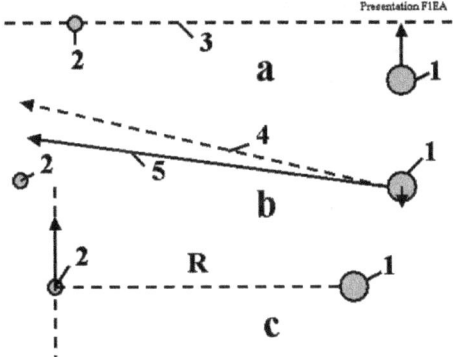

Fig. 8.1. Preparing for employment of the asteroid. Notations: 1 – space ship, 2 – asteroid, 3 – plane of maneuver. 4 – old ship direction, 5 – corrected ship direction. a) Reaching the plane of maneuver; b) Correcting the flight direction and reaching the requested radius; c) Connection to the asteroid.

When the apparatus is at the shortest distance R from the space body, it connects to the space body means of the net (Fig. 8.2a) or by the anchor (Fig. 8.2b) and rope. The apparatus rotates around the common center of gravity at the angle $j\varphi$ with angular speed $w\omega$ and linear speed $D\Delta V$. The cardioids of additional speed and direction of the apparatus are shown in [1] Fig. 11.4 (right side). The maximum additional velocity is $D\Delta V = 2V_a$, where V_a is the relative asteroid velocity when the coordinate center is located in the apparatus.

Fig. 8.2a shows how a net can be used to catch a small asteroid or meteorite. The net is positioned in the trajectory of the meteorite or small asteroid, supported in an open position by the inflatable ring and connected to the space apparatus by the rope. The net catches the asteroid and transfers its kinetic energy to the space apparatus. The space apparatus changes its trajectory and speed and then disconnects from the asteroid and spools the cable. If the asteroid is large, the astronaut team can use the asteroid anchor (Figs. 8.2b).

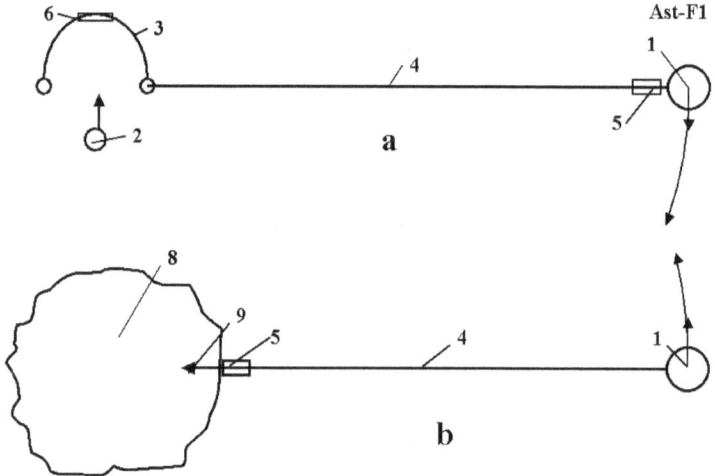

Fig. 8.2. a) Catching a small asteroid using net; b) Connection to a big asteroid using an anchor and cable. Notation: 1 – space ship, 2, 8 – asteroid, 3 – net with inflatable ring, 4 – cable (rope), 5 – load cabin, 6 – valve, 9 – anchor.

The astronauts use the launcher (a gun or a rocket engine) to fire the anchor (harpoon fork) into the asteroid. The anchor is connected to the rope and spool. The anchor is implanted into the asteroid and connects the space apparatus to the asteroid. The anchor contains the rope spool and a disconnect mechanism. The space apparatus contains a spool for the rope, motor, gear transmission, brake, and controller. The apparatus may also have a container for delivering a load to the asteroid and back (Fig. 8.2b). One possible design of the space anchor is shown in [1, Fig. 11.3]. The anchor has a body, a rope, a cumulative charge (shared charge), the rocket impulse (explosive) engine, the rope spool and the rope keeper. When the anchor strikes the asteroid surface the cumulative charge burns a deep hole in the asteroid and the rocket-impulse engine hammers the anchor body into the asteroid. The anchor body pegs the catchers into the walls of the hole and the anchor's strength keeps it attached to the asteroid. When the apparatus is to be disconnected from the asteroid, a signal is given to the disconnect mechanism.

If the asteroid is rotated with angular speed $w\omega$, its rotational energy can be used for increasing the velocity and changing the trajectory of the space apparatus. The rotary asteroid spools the rope on its body. The length of the rope is decreased, but the apparatus speed is increased (see a momentum theory in physics).

The ship can change the length of the cable. When the radius decreases, the linear speed of the apparatus increases; conversely, when the radius increases the apparatus speed decreases. The apparatus can obtain energy from the asteroid by increasing the length of the rope.

The computations and estimations show the possibility of making this method a reality in a short period of time.

An abandoned space vehicle or large piece of space debris in Earth orbit can also be used to increase the speed of the new vehicle and to remove the abandoned vehicle or debris from orbit.

9. Multi-reflex Propulsion Systems for Space and Air Vehicles and Energy Transfer for Long Distance*

The purpose of this article is to draw attention to the revolutionary idea of light multi-reflection. This idea allows the design of new engines, space and air propulsion systems, storage systems (for a beam or solar energy), transmission of energy (over millions of kilometers), creation of new weapons, etc. This method and the main innovations were offered by the author in 1983 in the former USSR. Now the author shows the immense possibilities of this idea in many fields of engineering – astronautics, aviation, energy, optics, direct conversion of light (laser beam) energy to mechanical energy (light engine), to name a few. This chapter considers the multi-reflex propulsion systems for space and air vehicles and energy transmission over long distances in space.

----

* A detailed manuscript was published by A.A. Bolonkin, *JBIS*, Vol. 57, No. 11/12. 2004, pp. 379–390, 2004. See also [1, Ch.12, pp.223-244].

Introduction

The reflection of light is the most efficient method to use for a propulsion system. It gives the maximum possible specific impulse (light speed is $3 \cdot 10^8$ m/s). The system does not expend mass. However, the light intensity in full reflection is very small, about 0.6×10^{-6} kg/ kW. In 1983 the author suggested the idea of increasing the light intensity by a multi-reflex method (multiple reflection of the light beam) and he offered some innovations to dramatically decrease the losses in mirror reflection (including a cell mirror and reflection by a super–conducting material). This allows the system to make some millions of reflections and to gain some Newtons of thrust per kW of beam power. This allows for the design of many important devices (in particular, beam engines [1, Ch.12[7]]) which convert light directly into mechanical energy and solve many problems in aviation, space, energy and energy transmission.

Description of innovation

Multi-reflex launch installation of a space vehicle. In a multiple reflection propulsion system a set of tasks appear: how to increase a mirror's reflectivity, how to decrease the light dispersion (from mirror imperfections and non-parallel surfaces), how to decrease the beam divergence, how to inject the beam between the mirrors (while keeping the light between the mirrors for as long as possible), how to decrease the attenuation (a mirror, prism material, etc), how to increase the beam range, and how much force the system has.

To solve of these problems, the author proposes [1] Ch.12[5], a special "cell mirror" which is very reflective and reflects light in the same direction from which it came, a "laser ring" which decreases the beam divergence, "light locks" which allows the light beam to enter but keep it from exiting, a "beam transfer", a "focusing prismatic thin lens", prisms, a set of lenses, mirrors located in space, on asteroids, moons, satellites, and so on.

Cell mirrors. To achieve the maximum reflectance, reduce light absorption, and preserve beam direction the author uses special *cell mirrors* which have millions of small 45° degree prisms (1 in Fig. 9.1a,g). Cell mirror are retroreflector cells or cube corner cells. A light ray incident on a cell is returned parallel to itself after three reflections (Fig. 9.1g). In the mirror, provided the refractive index of the prism is greater than $\sqrt{2}$ ($\cong 1.414$), the light will be reflected by total internal reflection. The small losses may be only from prism (medium) attenuation, scattering, or due to small surface imperfections and Fresnel reflections at the entrance and exit faces. Fresnel reflections do not result losses when the beam is perpendicular to the entry surface. No entry losses occur where the beam is polarized in parallel of the entry surface or the entry surface has an anti-reflection coating with reflective index $n_1 = \sqrt{n_0 n_2}$. Here n_0, n_2 are reflective indexes of the vacuum and prism respectively. These cell mirrors turn a beam (light) exactly back at 180° if the beam deviation is less 5–10° from a perpendicular to the mirror surface. For incident angles greater than $\sin^{-1}(n_1/n_2)$, no light is transmitted, an effect called total internal reflection. Here n is the refractive index of the medium and the lens ($n \approx 1$–4). Total internal reflection is used for our reflector, which contains two plates (mirrors) with a set of small corner cube prisms reflecting the beam from one side (mirror) to the other side (mirror) (Fig. 9.1b,c, f). Each plate can contain millions of small (30–100 μm) prisms from highly efficient optic material used in optical cables [1] Ch.12[9]. For this purpose a superconductivity mirror [1] Ch.12[5] may also be used,

Laser ring. The small lasers are located in a round ring (Fig. 9.1c). A round set of lasers allows us to increase the aperture, resulting in a smaller divergence angle θ. The entering round beam (9 in Fig. 9.1a) has slip θ (or $\theta/2$) to the vertical. The beam is reflected millions of times as is shown in Fig. 9.1b,c and creates a repulsive force F. This force may be very high, tens of N/kW (see the computation below) for motionless plates. In a vacuum it is limited only by the absorption (dB) of the prism material (see below) and beam divergence. For the mobile mirror (as for a launch vehicle) the wavelength increases and beam energy decreases as the mirrors move apart.

This system [1] Ch.12[5] can be applied to a space vehicle launch on a planet that has no atmosphere and small gravity (for example, the Moon; high gravity requires high beam power).

The offered multi-reflex light launcher, space and air focused energy transfer system is very simple (needing only special mirrors, lenses and prisms), and it has a high efficiency. One can directly transfer the light beam into space acceleration and mechanical energy. A distant propulsion system can obtain its energy from the Earth. However, we need very powerful lasers. Sooner or later the industry will create these powerful lasers (and cell mirrors) and the ideas presented here will become possible. The research on these problems should be started now.

Multi-reflex engines7 may be used in aviation as the energy can be transferred from the power stations on the ground to the aircraft using laser beams. The aircraft would no longer carry fuel and the engine would be lighter in weight so its load capability would double. The industry produces a one Megawatt (1000 kW) laser now. This is the right size for mid-weight aircraft (10–12 tons).

The linear light engine does not have a limit to its speed and may be used to launch space equipment and space ships in non-rockets method described in [1] Ch.12[10-29]. This method is certain also to have many military applications.

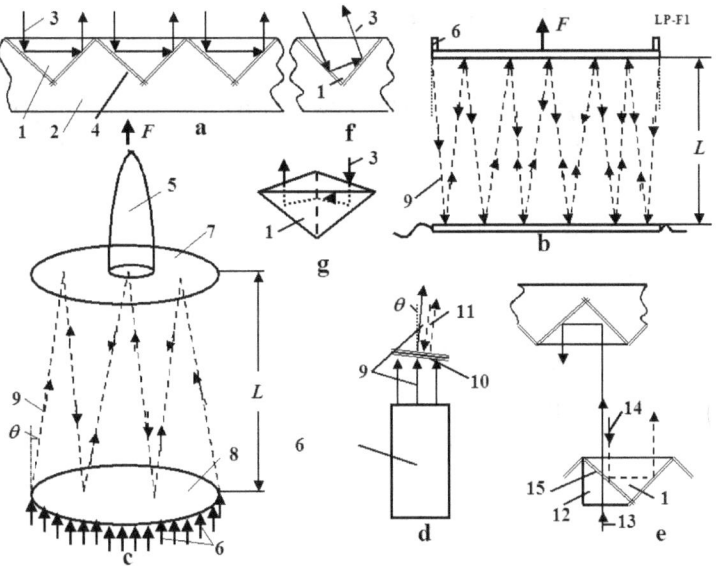

Fig. 9.1. Space launcher. Notations are: 1 – prism, 2 – mirror base, 3 – laser beam, 4 – mirror after chink (optional), 5 – space vehicle, 6 – lasers (ring set of lasers), 7 – vehicle (ship) mirror, 8 – planet mirror, 9 – laser beam, 10 – multi-layer dielectric mirror, 11 – laser beam after multi-reflection (wavelength $\lambda_{11} > \lambda_9$), 12 – additional prism, 13 – entry beam, 14 – return beam, 15 – variable chink between main and additional prisms. (a) Prism (cell, corner cube) reflector. (b) Beam multi-reflection, (c) Launching by multi-reflection, (d) The first design of the light lock, (e) The second design of the light lock, (f) Reflection in the same direction when the beam is not perpendicular to mirror surface, (g) Mirror cell (retroreflector cell or cube corner cell). A light ray incident on it is returned parallel to itself after three reflections.

10. Electrostatic Solar wind propulsion*

A revolutionary method for space flights in outer space is suggested by the author. Research is present to shows that an open high charged (100 MV/m) ball of small diameter (4–10 m) made from thin film collects solar wind (protons) from a large area (hundreds of square kilometers). The proposed propulsion system creates many Newtons of thrust, and accelerates a 100 kg space probe up to 60–100 km/s for 100–800 days. The 100 kg space apparatus offers flights into Mars orbit of about 70 days, to Jupiter about 150 days, to Saturn about 250 days, to Uranus about 450 days, to Neptune about 650 days, and to Pluto about 850 days.

The author has computed the amount thrust (drag), to mass of the charged ball, and the energy needed for initial charging of the ball and discusses the ball discharging in the space environment. He also reviews apparent errors found in other articles on these topics. Computations are made for space probes with a useful mass of 100 kg.

----

*The work was presented as AIAA-2005-3857 at the 41st Propulsion Conference, 10–13 July 2005, Tucson, Arizona, USA. See also [1, Ch.13, pp.245-270].

Introduction.
Brief information about solar Wind. The Sun emits plasma which is a continuous outward

flow (solar Wind) of ionized solar gas throughout our solar system. The solar wind contains about 90% protons and electrons and some quantities of ionized α-particles and gases. It attains speeds in the range of 300–750 km/s and has a flow density of $5\times10^7 - 5\times10^8$ protons/electrons/cm²s. The observed speed rises systematically from low values a 300–400 km/s to high values of 650–700 km/s in 1 or 2 days and then returns to low values during the next 3 to 5 days ([1], Fig. 13.1). Each of these high-speed streams tends to appeal at approximately 27-day intervals or to recur with the rotation period of the Sun. On days of high Sun activity the solar wind speed reaches 1000 (and more) km/s and its flow density $10^9 - 10^{10}$ protons/electrons/ cm²s, 8–70 particles in cm³. The Sun has high activity periods some days each year.

The pressure of the solar wind is very small. For full braking it is in the interval $2.5\times10^{-10} \div 6.3\times10^{-9}$ N/m². This value is double when the particles have full reflection. The interstellar medium also has high energy particles. Their density is about 1 particle/cm³.

Brief description of the propulsion system.

Space propulsion system. The suggested propulsion system is very simple. It includes a hollow ball made up of a thin, strong, film – covered conductive layer or a ball of thin net. The ball is charged by high voltage static electricity which creates a powerful electrostatic field around it. Charged particles of solar wind of like charges repel and particles with the unlike charges attract. A small proportion of them run through the ball, a larger proportion flow round the ball in hyperbolic trajectory into the opposive direction, and another proportion are deviates from their initial direction in hyperbolic curves. As a result the charged ball has drag when the ball speed is different from solar speed (Fig.10.1). The drag also occurs when the particles and the ball have the same electrical charge. In this case the particles are repelled from the charged ball (Fig. 10.1) and brake it. This solar wind drag provides thrust in our proposed propulsion system. The pressure of solar wind is very small, but the offered system (a charged ball of radius 6–10 m) collects particles (protons or electrons) from a large area (an area of tens of kilometers radius for protons and hundreds of kilometers for electrons), creates a thrust of some Newtons and a 100-kg space ship reaches speeds of tens of km/s in 50–300 days (see theory and computation below and References [1] Ch.13[29, 42–47]).

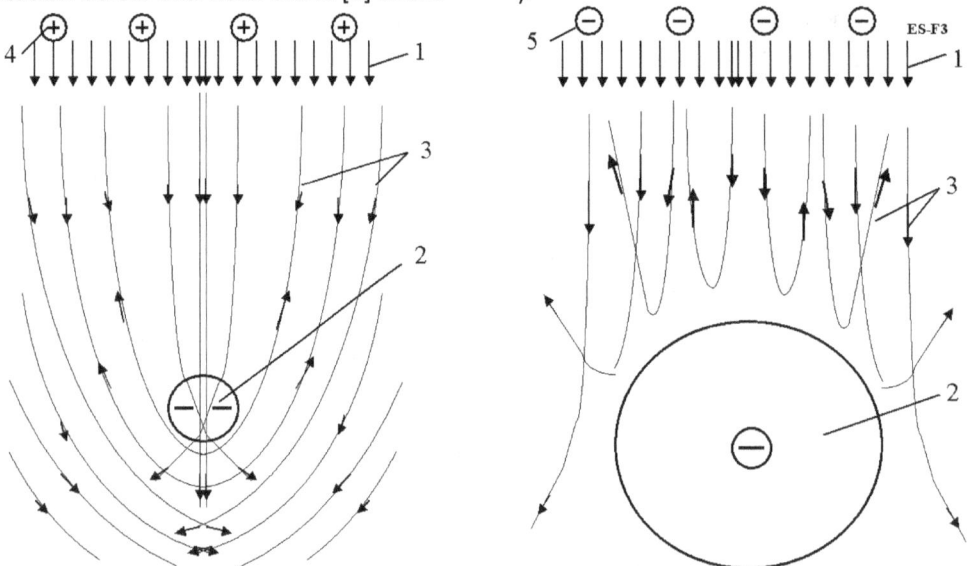

Fig. 10.1 (Left) Hyperbolic trajectory of protons around static negative charge (or electrons around positive charge) (unlike charges). Notations are: 1 – solar wind (charged particles); 2 – hollow negatively charged ball of thin film; 3 – hyperbolic trajectory of charged particles; 4 – positively charged particles (protons).

(Right). Trajectory of particles having the same charge as the ball. Notations are: 5 – negatively charged particles.

The proposed new propulsion mechanism differs from previous concepts in very important respects; including the coupling to the protons of the solar wind using an open single-charge ball. The opposite charge is expelled into infinite space. This innovation increases the area of influence by up to hundreds of kilometers for protons and allows the acquisition of significant vehicle thrust. This thrust is enough to accelerate a heavy space craft to very high speed and permits very short flight times to far planets.

The offered revolutionary propulsion system has a simple design, which can give useful acceleration to various types of spacecraft. The offered propulsion system creates many Newtons of thrust, and can accelerate a 100 kg space probe up to 60–100 km/sec in 100–800 days.

In the offered wind propulsion system the particles run away from the ball, brake and return in infinity for initial speed. These premises must be examined using more complex theories to account for the full intersection between the suggested installation and solar wind (including thermonuclear reactions). This would be a revolutionary breakthrough in interplanetary space exploration.

The author has developed the initial theory and the initial computations to show the possibility of the offered concept. He calls on scientists, engineers, space organizations, and companies to research and develop the offered perspective concepts.

11. Electrostatic Utilization of Asteroids for Space Flight*

The author offers an electrostatic method for changing the trajectory of space probes. The method uses electrostatic force and the kinetic or rotary energy of asteroids, comet nuclei, meteorites or other space bodies (small planets, natural planet satellites such a moons, space debris, etc.) to increase or decrease ship/ probe speed by 1000 m/s or more and to achieve any new direction in outer space. The flight possibilities of spaceships and probes are thereby increased by a factor of millions.

----

*The full text was presented by the author as Paper AIAA-2005-4032 at the 41 Propulsion Conference, 10–12 July 2005, Tucson, Arizona, USA; or [1, Ch.14, pp.271-280].

Description

. The method includes the following main steps (Fig. 11.1):

(a) Finding an asteroids using a locator or telescope (or looking in a catalog) an asteroid and determining its main parameters (location, mass, speed, direction, rotation); selecting the appropriate asteroid; computing the required position of the ship with respect to the asteroid.

(b) Correcting the ship's trajectory to obtain the required position; convergence of the ship with the asteroid.

(c) Charging the asteroid and space apparatus ball using a charge gun.

(d) Obtaining the necessary apparatus position and speed for the apparatus by flying it around the space body and changing the charge of the apparatus and space body (asteroids).

(e) Discharging the space apparatus and the space body.

The equipment requires for changing a probe (spacecraft) trajectory includes:

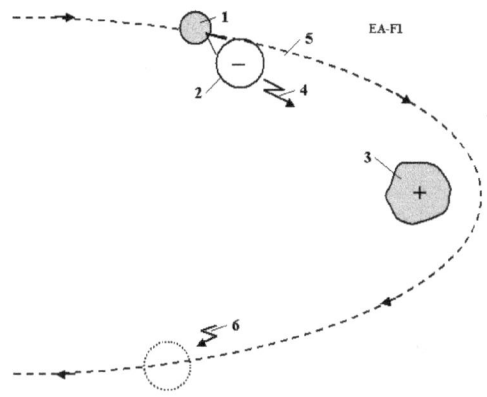

Fig. 11.1. Method of electrostatic maneuvers of the space apparatus. Notations: 1 – space apparatus, 2 – charged ball, 3 – asteroid, 4 – charged gun, 5 – new apparatus trajectory, 6 – discharging the apparatus and asteroid.

(a) A charging gun.

(b) Devices for finding and measuring the asteroids (space bodies), and computing the trajectory of the spacecraft relative of the space body.

(c) Devices for spacecraft guidance and control.

(d) A device for discharging of the apparatus and asteroids (space body) (see [1] Fig. 14.1).

12. Electrostatic Levitation on the Earth and Artificial Gravity for Space Ships and Asteroids*

The author offers and researches the conditions which allow people and vehicles to levitate on the Earth using the electrostatic repulsive force. He shows that by using small electrically charged balls, people and cars can take flight in the atmosphere. Also, a levitated train can attain high speeds. He has computed some projects and discusses the problems which can appear in the practical development of this method. It is also shown how this method may be used for creating artificial gravity (attraction force) into and out of space ships, space hotels, asteroids, and small planets which have little gravity.

----

*Presented as paper AIAA-2005-4465 at 41 Propulsion Conference, 10–13 July 2005, Tucson, Arizona, USA; or [1, Ch. 15, pp. 281-302].

Brief description of innovation

It is known that like electric charges repel, and unlike electric charges attract (Fig. 12.1a,b,c). A large electric charge (for example, positive) located at altitude induces the opposite (negative) electric charge at the Earth's surface (Figs. 12.1d,e,f,g) because the Earth is an electrical conductor. Between the upper and lower charges there is an electric field. If a small negative electric charge is placed in this electric field, this charge will be repelled from the like charges (on the Earth's surface) and attracted to the upper charge (Fig. 12.1d). That is the electrostatic lift force. The majority of the lift force is determined by the Earth's charges because the small charges are conventionally located near the Earth's surface. As shown below, these small charges can be connected to a man or a car and have enough force to lift and supports them in the air.

The upper charge may be located on a column as shown in Fig. 12.1d,e,f,g or a tethered air balloon (if we want to create levitation in a small town) (Fig. 12.1e), or air tube (if we want to build a big highway), or a tube suspended on columns (Fig. 12.1f,g). In particular, the charges may be at two identically charged plates, used for a non-contact train.

A lifting charge may use charged balls. If a thin film ball with maximum electrical intensity of below 3×10^6 V/m is used, the ball will have a radius of about 1 m (the man mass is 100 kg). For a 1 ton car, the ball will have a radius of about 3 m (see the computation in [1] and Fig. 15.2g,h,i). If a higher electric intensity is used, the balls can be small and located underneath clothes.

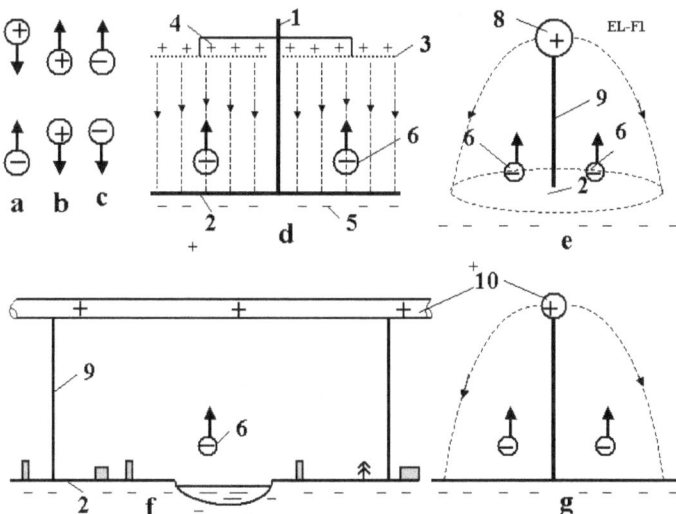

Fig. 12.1. Explanation of electrostatic levitation: a) Attraction of unlike charges; b,c) repulsion of like charges; d) Creation of the homogeneous electric field (highway); e) Electrical field from a large spherical charge ; f,g) Electrical field from a tube (highway) (side and front views). Notations are: 1, 9 – column, 2 – Earth (or other) surface charged by induction, 3 – net, 4 – upper charges, 5 – lower charges, 6 – levitation apparatus, 8 – charged air balloon, 9 – column, 10 – charged tube.

13. Guided Solar Sail and Energy Generator*

A solar sail is a large thin film mirror that uses solar energy for propulsion. The author proposed innovations and a new design of Solar sail in 1985 [1] Ch.161. This innovation allows (main advantages only):
1) An easily controlled amount and direction of thrust without turning a gigantic sail;
2) Utilization of the solar sail as a power generator (for example, electricity generator);
3) Use of the solar sail for long-distance communication systems.

---

* The detail manuscript was presented as AIAA-2005-3857 on the 41st Propulsion Conference, 10–12 July 2005, Tucson, Arizona, USA; or see [1, Ch.16, pp. 303-308].

Description of innovation and their advantages

The proposed innovation of a solar sail[1] is presented in Fig. 13.1. Theory developed in author publication [1, Ch.16[2]] may be useful for flight analysis. The solar sail contains: a space ship, 1, a spherical reflector, 2, a mirror, 5, and additional devices to support spherical reflector, control the thrust direction, and convert the light energy into electricity and additional thrust. Emergy in the proposed sail can increase the thrust over time.

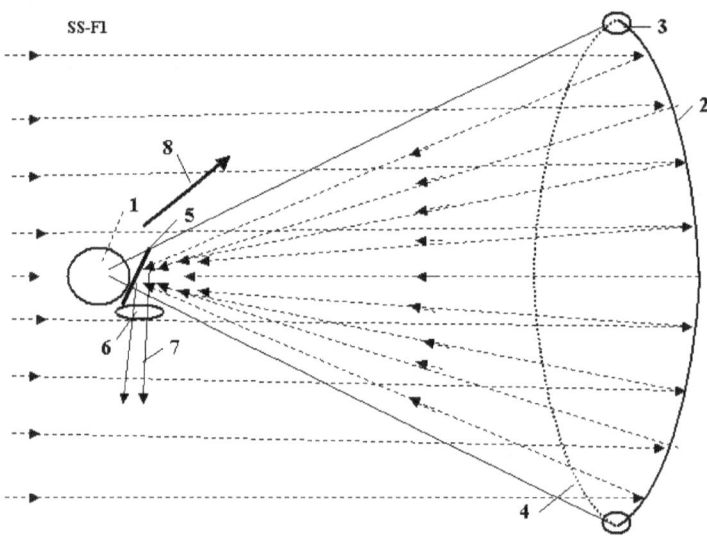

Fig. 13.1. Proposed guided solar sail and electricity generator. Notations are: 1 – space ship; 2 – thin film reflector; 3 – inflatable (or electrically charged) toroid which support the reflector in an open (unfolded) position; 4 – transparent thin film or light charged net which support the spherical form of the reflector; 5 – control mirror, which guides thrust direction; 6 – lens or trap for communication; 7 – reflected beam (located at the center of the ship's mass). 8 – direction of thrust.

The suggested propulsion system works in the following way. The reflector, 2, focuses the sunlight into the control ship mirror, 5, located at the spaceship's center of mass. We are able to change the position of this mirror, send the focused beam in the right direction and achieve the necessary thrust direction without turning the space sail because the space sail is large, turning it is very complex problem, but this problem is avoided in the suggested design.

If we direct the solar beam into the ship, we can convert the huge solar energy into any other sort of energy, for example, into electricity using a conventional method (solar cell or heat machine). A reflector of 100×100 m^2 produces 14,000 kW energy at 1 AU. The developed ion thrusters are very efficient and have a high specific impulse, but they need a great amount of energy. We have this

The offered system can be also used for long distance communication by sending a focused beam is to the Earth and transmitting the necessary information.

The author has also proposed a method using surface tension of a solar sail and a solar mirror [1, Ch.16[10]].

The suggested revolutionary propulsion system uses current technology and may be produced in the near future but needs detailed research and computation.

14. Radioisotope Space Sail and Electro-Generator*

Radioisotope sail is a thin film of an alpha particle emitting radioisotope deposited on the back of a plastic sail that can provide useful quantities of both propulsion and electrical power to a deep space vehicle. The momentum kick of the emitted alpha particles provides radioisotope sail thrust levels per square meter comparable to that of a solar sail at one astronautical unity (1 AU). The electrical power generated per 1 square m is comparable to that obtained from solar cells at 1 AU. Radioisotope sail systems will maintain these propulsion and power levels at distances from the Sun where solar powered systems are ineffective.

The propulsion and power levels available from this simple and reliable high-energy-density system would be useful for supplying propulsion and electrical power to a robotic deep space mission to the Oort Cloud or beyond, or to a robotic interstellar flyby or rendezvous probe after its arrival at the target star.

----

* Detailed work was presented by the author as AIAA-2005-4225 at the 41st Propulsion Conference, 10–12 July, 2005, Tucson, Arizona, USA; or see [1, Ch.17, pp.309-316].

Description of method and innovations
Brief history of innovation

The idea of using a radioisotope recoil propulsion as it is shown in Fig.14.1a is very old [1, Ch.17[15]]. The author has proposed many innovations in method is using radioisotope space sail and electric generators in patent applications [1, Ch.17[1-13]] in 1983 and in paper IAF 92-0573 presented to the World Space Congress in 1992]1, Ch.17[14]]. The work [1, Ch.17[16]] written in 1995 summarized the knowledge for the conventional case in Fig. 14.1a. Bolonkin innovations decrease the weight of traditional radioisotope sail (RadSail, RS, IsoSail) by 2–4 times; increase the thrust by 2–3 times, and the electric power by 2 times and allows control of thrust and thrust direction without needing to turn the large RadSail.

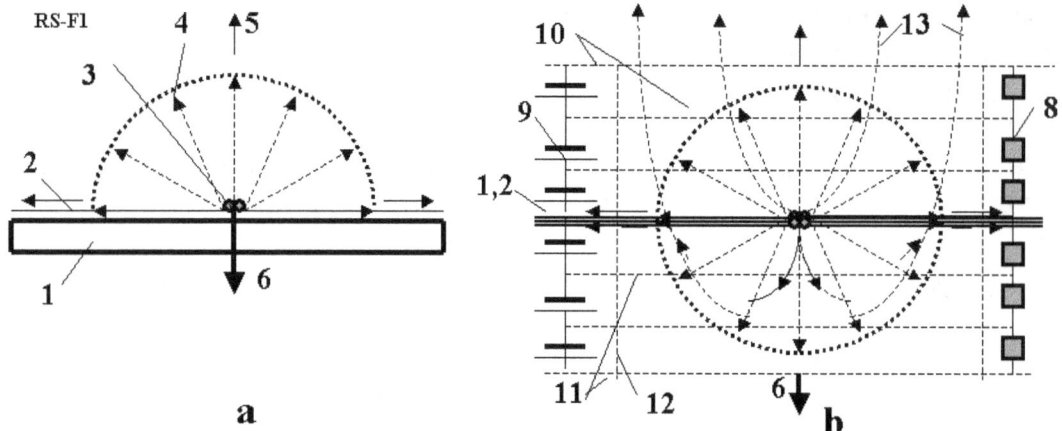

Fig. 14.1. a) Conventional radioisotope sail; b) suggested (innovated) radioisotope sail. Notations are: 1 – substrate (base of sail), 2 – isotope layer, 3 – isotope atom, 4 – alpha particles, 5 – direction of 1/6 particle flow, 6 – thrust, 8 – electric loading, 9 – initial charging, 10, 11, 12 – condenser nets, 13 – particle trajectory.

Offered innovation allows us to reach a probe speed of more than 2000 km/s, so the design may be used for interstellar probes.

This method allows nuclear waste and unnecessary nuclear bombs to be used for producing the radioisotope material.

The offered method is realistic at the present time, has a high possibility to being successful, and is much cheaper for deep missions than other currently proposed method.

15. Electrostatic Solar Sail*

The solar sail has become well-known after much discussion in the scientific literature as a thin continuous plastic film, covered by sunlight-reflecting appliquéd aluminum. Earlier, there were attempts to launch and operate solar sails in near-Earth space and there are experimental projects planned for long powered space voyages. However, as currently envisioned, the solar sail has essential disadvantages. Solar light pressure in space is very low and consequently the solar sail has to be very large in area. Also it is difficult to unfold and unfurl the solar sail in space. In addition it is necessary to have a rigid framework to support the thin material. Such frameworks usually have great mass and, therefore, the spacecraft's acceleration is small.

Here, the author proposes to discard standard solar sail technology (continuous plastic aluminum-coated film) with the intention instead of using millions of small, very thin aluminum charged plates and to release these plates from a spacecraft, instigated by an electrostatic field. Using this new technology, the solar sail composed of millions of plates can be made gigantic area but have very low mass. The acceleration of this new kind of solar sail may be as much as 300 times that achieves by an ordinary solar sail. The electrostatic solar sail can even reach a speed of about 300 km/s (in a special maneuver up to 600–800 km/sec). The electrostatic solar sail may be used to move a large spaceship or to act as an artificial Moon illuminating a huge region of the Earth's surface.

---
*See [1, Ch.18, pp. 317-326].

Brief description of the innovations

A conventional solar sail is a dielectric thin film (thickness 5 mkm = 5000 nm) with an aluminum layer 100 nm thick, and it has 90% reflectivity. The weight of one square meter is 5–7 g/m². If it accelerates by itself the maximum acceleration is about 1 mm/s². However, the gigantic thin film needs a rigid structure to support the very thin film in an unfolded position and to unable it to be controlled. This rigid structure has a large weight, so it is very difficult to launch and to unfurl the structure in space. All attempts to do this (for example, to unfurl the inflatable radio-antennas in space) have failed.

The author proposes to use small thin charged aluminums plates (petals) supported by a central electrostatic ball and rotated around the ball (Fig. 15.1). They rotate also around their own axis and main thin a direction perpendicular to the solar rays. The diameter of the plate-petals is small, about 1 mm or less, and, it is not a necessity to use the dielectric film. The aluminum film may be very thin because the individual petal size is small.

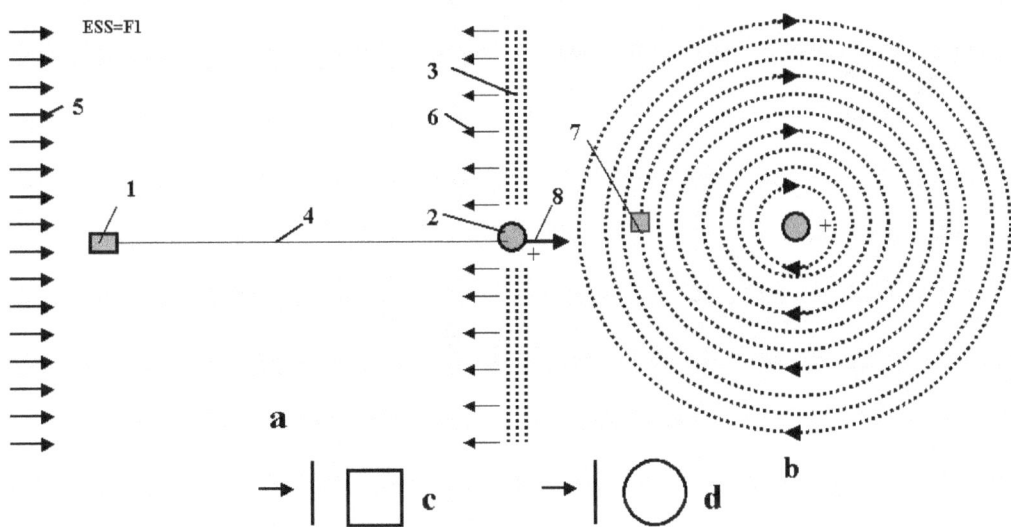

Fig. 15.1. The proposed electrostatic solar sail. a. Side view; b. Front view; c. Side and front views of square petal; d. Side and front views of round petal. Notation: 1 – spaceship, 2 – charged ball, 3 – charged plate-petals, 4 – cable connecting the ship and the ball, 5 – solar rays, 6 – reflected rays, 7 – charged petals, 8 – thrust (drag).

16. Recombination Space Jet Propulsion Engine*

There are four known ionized layers in the Earth's atmosphere, located at an altitude of 85–400 km. Here the concentration of ions reaches millions of particles in 1 cubic centimeter. In the inter-planetary medium the concentration of ion reaches $10–10^3$ particles in 1 cm³ and in interstellar space it is about 1–10 in 1 cm³. As a result there is interaction between solar radiation in the Earth's atmosphere, solar wind, and galactic radiation.

About 90% these particles are protons and electrons. The particle density is low and they can exist for a long time before they come into collision with each other. However, if we increase the density of the particles in an engine, they collide with one another, recombine, warm up, leave the propulsion system with high speed, and create thrust.

The energy of recombination is significantly more than the heat capability of conventional fuel and the specific impulse of the propulsion system is high.

The author proposes collecting and concentrating charged particles from a large area using a magnetic field. Space ships, space apparatus, and satellites would then not need fuel and could be accelerated or fly to infinity. This may be a revolution in aerospace.

========
*See [1, Ch. 19, pp. 327-338].

Description of innovation

In the recombination propulsion engine contains a tube with an intake and a nozzle, and a solenoid (Fig. 16.1).

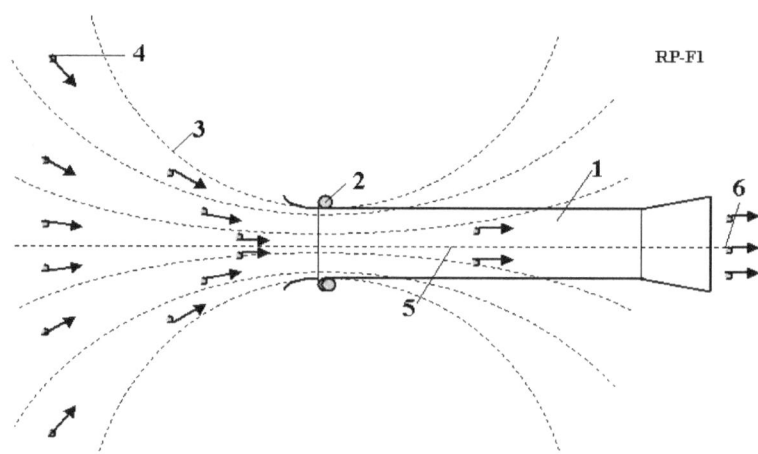

Fig. 16.1. Recombination space jet propulsion engine (actuator of magnetic field). Notations are: 1 engine, 2 – solenoid, 3 – magnetic lines, 4 – charged particles, 5 – recombination zone, 6 – exit.

The solenoid may be conventional or superconductive. It produces a powerful magnetic field, which collects charged particles. If the density of the charged particles is sufficient (the distance the particles travel is less than the tube length) the particles came into contact with each other and recombine.

This minimum energy is more than the energy of the most efficient chemical reaction, $H_2 + O = H_2O$, by hundreds of times. This means the specific impulse of the recombination engine will be very high. The heating of engine walls will be small, however, because the density of the particle gas is low. Using this proposed method, we do not need to expend fuel and can achieve a large acceleration of a space vehicle, or support the satellite at altitude for an infinite amount of time.

Idea's are needed in research and development of this method.

17. Electronic Sail*

A solar sail reflects solar light and can be a used as propulsion system, as described in [1, Ch. 16]. It needs thin film of a very large area. This manuscript proposes a new way of creating a reflecting surface of large area using an electronic method. This method needs research and development but it may be easier and more efficient than the film method.

----

*See [1, Ch.19, pp.334-335].

Brief description of innovation

The proposed electronic sail has a positive charge, 1 (see Fig. 17.1). The free electrons, 2, are injected into space around the positive charge so they rotate around the center of the charge and form a thin disk in a plane perpendicular to the direction of the Sun light. If the concentration of electrons is sufficient, they will reflect the solar light like a mirror and produce thrust.

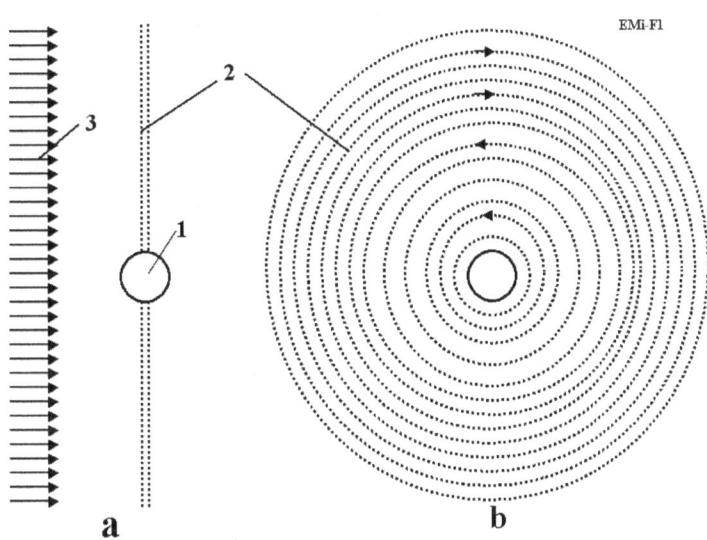

Fig. 17.1. Electronic solar sail. **a** – side view, **b** – front view. Notations are: 1 – positive charge, 2 – electronic disk, 3 – solar light.

This electronic sail may be an electrostatic solar wind sail, as described in [1] Chapter 13, if the central charge is positive. The solar wind electrons became concentrated around it and the mass of electrons reflects the solar light. Thrust from the solar wind is small because the electron mass is about 2000 times less than the proton mass, but the solar light pressure is thousands of times greater than solar wind (protons) pressure. The offered installation may also be used as a space mirror to illuminate the Earth's surface. This idea needs further research.

References

Many works noted below the reader can find on site Cornel University <http://arxiv.org/>, sites <http://www.vixra.org>, conferences AIAA, <http://aiaa.org/>, search "Bolonkin")

1. Bolonkin A.A., (2005) "Non-Rocket Space Launch and Flight", Elsevier. 2005, http://www.archive.org/details/Non-rocketSpaceLaunchAndFlight,
2. K.E. Tsiolkovski:"Speculations Abot Earth and Sky on Vesta", Moscow, Izd-vo AN SSSR, 1959; Grezi o zemle i nebe (in Russian), Academy of Sciences, USSR., Moscow, p. 35, 1999.
3. Geoffrey A. Landis, Craig Cafarelli, The Tsiolkovski Tower Re-Examined, *JBIS*, Vol. 32, p. 176–180, 1999.
4. Y. Artsutanov. Space Elevator, http://www.liftport.com/files/Artsutanov_Pravda_SE.pdf.
5. A.C. Clarke: *Fountains of Paradise*, Harcourt Brace Jovanovich, New York, 1978.
6. Bolonkin A.A. (2006), Book "*New Concepts, Ideas and Innovation in Aerospace*", NOVA, 2008. http://www.archive.org/details/NewConceptsIfeasAndInnovationsInAerospaceTechnologyAndHumanSciences
7. Bolonkin A.A. (2007), "Macro-Engineering: Environment and Technology", NOVA, 2008. http://Bolonkin.narod.ru/p65.htm, http://www.archive.org/details/Macro-projectsEnvironmentsAndTechnologies
8. Bolonkin A.A. (2008), "New Technologies and Revolutionary Projects", Scribd, 2008, 324 pgs, http://www.archive.org/details/NewTechnologiesAndRevolutionaryProjects
9. Bolonkin A.A., Book "Non-Rocket Space Launch and Flight", Elsevier. 2006, Ch. 9 "*Kinetic Anti-Gravotator*", pp. 165-186; http://Bolonkin.narod.ru/p65.htm, http://www.scribd.com/doc/24056182 ; Main idea of this Chapter was presented as papers COSPAR-02, C1.1-0035-02 and IAC-02-IAA.1.3.03, 53rd International Astronautical Congress. The World Space Congress-2002, 10-19 October 2002, Houston, TX, USA, and the full manuscript was accepted as AIAA-2005-4504, 41st Propulsion Conference, 10-12 July 2005, Tucson, AZ, USA.http://aiaa.org search "Bolonkin".
10. Bolonkin A.A., Book "Non-Rocket Space Launch and Flight", Elsevier. 2006, Ch.5 "*Kinetic Space Towers*", pp. 107-124, Springer, 2006. http://Bolonkin.narod.ru/p65.htm or . http://www.archive.org/details/Non-rocketSpaceLaunchAndFlight
11. Bolonkin A.A., "Transport System for Delivery Tourists at Altitude 140 km", manuscript was presented as paper IAC-02-IAA.1.3.03 at the World Space Congress-2002, 10-19 October, Houston, TX, USA. http://Bolonkin.narod.ru/p65.htm ,
12. Bolonkin A.A. (2003), "Centrifugal Keeper for Space Station and Satellites", JBIS, Vol.56, No. 9/10, 2003, pp. 314-327. http://Bolonkin.narod.ru/p65.htm . See also [10] Ch.3.
13. Bolonkin A.A., Book "Non-Rocket Space Launch and Flight", Elsevier. 2006, Ch.3 "*Circle Launcher and Space Keeper*", pp.59-82. , http://www.archive.org/details/Non-rocketSpaceLaunchAndFlight
14. Bolonkin A.A., Book "New Concepts, Ideas and Innovation in Aerospace", NOVA, 2008, Ch. 3 " *Electrostatic AB-Ramjet Space Propulsion)*", pp.33-66. http://www.archive.org/details/NewConceptsIfeasAndInnovationsInAerospaceTechnologyAndHumanSciences
15 Bolonkin A.A., Book "New Concepts, Ideas and Innovation in Aerospace", NOVA, 2008, Ch.12, pp.205-220, "*AB- Levitrons and Their Applications to Earth's Motionless Satellites*". http://www.archive.org/details/NewConceptsIfeasAndInnovationsInAerospaceTechnologyAndHumanSciences
16. Bolonkin A.A., Book "Macro-Projects: Environment and Technology", NOVA, 2008, Ch.10, pp.199-226, " *AB-Space Propulsion*", http://Bolonkin.narod.ru/p65.htm . http://www.archive.org/details/Macro-projectsEnvironmentsAndTechnologies
17. Bolonkin A.A., Magnetic Suspended AB-Structures and Moveless Space Satellites. http://www.scribd.com/doc/25883886 .

18. Bolonkin A.A., Femtotechnology: Design of the Strongest AB-Matter for Aerospace. http://www.archive.org/details/FemtotechnologyDesignOfTheStrongestAb-matterForAerospace.
19. Bolonkin A.A., Converting of any Matter to Nuclear Energy by AB-Generator and Aerospace . http://www.archive.org/details/ConvertingOfAnyMatterToNuclearEnergyByAb-generatorAndAerospace.
20. Bolonkin A.A., LIFE. SCIENCE. FUTURE (Biography notes, researches and innovations). Scribd, 2010, 208 pgs. 16 Mb.
http://www.archive.org/details/Life.Science.Future.biographyNotesResearchesAndInnovations
21. Krinker M., Magnetic-Space-Launcher. http://www.scribd.com/doc/24051286/
22. Krinker M., Review of Space Towers. http://www.scribd.com/doc/26270139, http://arxiv.org/ftp/arxiv/papers/1002/1002.2405.pdf
23. Pensky O.G., Monograph "*Mathematical Models of Emotional Robots*", Perm, 2010, 193 ps (in English and Russian) (http://arxiv.org/ftp/arxiv/papers/1011/1011.1841.pdf).
24. Pensky O.G., K. V. Chernikov"*Fundamentals of Mathematical Theory of Emotional Robots* " (http://www.scribd.com/doc/40640088/).

25. Bolonkin A.A., Small Non-Expensive Electric Cumulative Thermonuclear Reactors. USA, Lulu,143 ps. ISBN 978-1-312-62280-7. http://viXra.org/abs/1703.0199
26. Bolonkin A.A., Preon Interaction Theory and Model of Universe. USA, Lulu.ISBN 978-1-365-80174-7. 102 ps. http://viXra.org/abs/1703.0200 , https://archive.org/download/BookPreonTheoryAndUniverseFromLulu3517
27. *Wikipedia.* Some background material in this article is gathered from Wikipedia under the Creative Commons license. http://wikipedia.org .

One design of aircraft XB-70

Space launcher Orion

Part B
Popular Reviews of New Consepts, Ideas, Innovation in Space Launch and Flight
Abstract

In the past years the author and other scientists have published a series of new methods which promise to revolutionize the space propulsion systems, space launching and flight. These include the electrostatic AB-ramjet space propulsion, beam space propulsion, MagSail, high speed AB-solar sail, transfer electricity in outer space, simplest AB-thermonuclear space propulsion, electrostatic linear engine and cable space launcher, AB-levitrons, electrostatic climber, AB-space propulsion, convertor any matter in nuclear energy, femtotechnology, wireless transfer of energy, magnetic space launcher, railgun, superconductivity rail gun, etc.

Some of them have the potential to decrease launch costs thousands of time, other allow to change the speed and direction of space apparatus without the spending of fuel.

The author reviews and summarizes some revolutionary propulsion systems for scientists, engineers, inventors, students and the public.

Key words: Non-rocket propulsion, non-rocket space launching, non-rocket space flight, electrostatic AB-ramjet space propulsion, beam space propulsion, MagSail, high speed AB-solar sail, transfer electricity in outer space, simplest AB-thermonuclear space propulsion, electrostatic linear engine and launcher, AB-levitrons, electrostatic climber, AB-space propulsion, convertor any matter in nuclear energy, femtotechnology, wireless transfer of energy, magnetic space launcher, railgun, superconductivity railgun.

Introduction

Brief history. Rockets for military and recreational uses date back to at least 13th century China. Modern rockets were born when Goddard attached a supersonic (de Laval) nozzle to a liquid-fueled rocket engine's combustion chamber. These nozzles turn the hot gas from the combustion chamber into a cooler, hypersonic, highly directed jet of gas, more than doubling the thrust and raising the engine efficiency from 2% to 64%. In 1926, Robert Goddard launched the world's first liquid-fueled rocket in Auburn, Massachusetts.

After World War 2 the missile systems have received the great progress and achieved a great success. But rockets are very expensive and have limited possibilities. In the beginning 21th century the researches of non-rocket launch and flight started [1],[5]-[22]. The non-rocket systems which promise to decrease the space launch and flight cost in hundreds times. Some of them are described in this review.

Current status of non-rocket space launch and flight systems. Over recent years interference-fit joining technology including the application of space methods has become important in the achievement of space propulsion system. Part results in the area of non-rocket space launch and flight methods have been patented recently or are patenting now.

Professor Bolonkin made a significant contribution to the study of the different types of non-rocket space launch and flight in recent years [1]-[22] (1982-2011). Some of them are presented in given review:

Electrostatic AB-ramjet space propulsion is researched in [2, Ch.2]; Beam space propulsion is described in [2, Ch. 3]; Magnetic Space Sail is presented in [2, Ch. 4]; High speed AB-solar sail is developed in [2, Ch.5]; Transfer electricity in outer space is offered in [2, Ch. 6]; Simplest AB-thermonuclear space propulsion is suggested in [2] Ch.7; Electrostatic linear engine and cable space launcher is presented in [2, Ch.10]; AB-levitrons are in [2, Ch. 12]; Electrostatic climber is researched in [3, Ch. 4]; AB-space propulsion is presented in [16] and [3, Ch.10]; Wireless transfer of energy is described in [4, Part A, Ch.3]; Magnetic space launcher is offered in [4, Part A, Ch.6]; Railgun Launch System is suggested in [4, Part A, Ch.7]; Superconductivity rail gun is presented in [6] and in [4, Part A, Ch.3]; Convertor any matter in nuclear energy and photon rocket is offered and researched in [4, Part A, Ch.1], [7], [19]; Femtotechnology and its application into aerospace technology is suggested and researched in [4, Part A, Ch.2], [8],[18]. Some of these system were developed in [9]-[23].

Significant scientific, interplanetary and industrial use did not occur until the 20th century, when rocketry was the enabling technology of the Space Age, including setting foot on the Moon.

But rockets are very expensive and have limited possibilities. In the beginning 21th century the researches of non-rocket launch and flight started [1], [5]-[8].Some of them are described in this review.

Main types of Non-Rocket Space Propulsion System

Contents:
1. Electrostatic AB-ramjet space propulsion,
2. Beam space propulsion,
3. MagSail,

4. High speed AB-solar sail,
5. Transfer electricity in outer space,
6. Simplest AB-thermonuclear space propulsion,
7. Electrostatic linear engine and cable space launcher,
8. AB-levitrons,
9. Electrostatic climber,
10. AB-space propulsion,
11. Wireless transfer of energy,
12. Magnetic space launcher,
13. Railgun,
14. Superconductivity rail gun.
15. Convertor any matter in nuclear energy and photon rocket,
16. Femtotechnology and its application into aerospace technology.

1. Electrostatic AB-ramjet space propulsion*

A new electrostatic ramjet space engine is proposed and analyzed. The upper atmosphere (85 - 1000 km) is extremely dense in ions (millions per cubic cm). The interplanetary medium contains positive protons from the solar wind. A charged ball collects the ions (protons) from the surrounding area and a special electric engine accelerates the ions to achieve thrust or decelerates the ions to achieve drag. The thrust may have a magnitude of several Newtons. If the ions are decelerated, the engine produces a drag and generates electrical energy. The theory of the new engine is developed. It is shown that the proposed engine driven by a solar battery (or other energy source) can not only support satellites in their orbit for a very long time but can also work as a launcher of space apparatus. The latter capability includes launch to high orbit, to the Moon, to far space, or to the Earth atmosphere (as a return thruster for space apparatus or as a killer of space debris). The proposed ramjet is very useful in interplanetary trips to far planets because it can simultaneously produce thrust or drag and large electric energy using the solar wind. Two scenarios, launch into the upper Earth atmosphere and an interplanetary trip, are simulated and the results illustrate the excellent possibilities of the new concept.

----
* Presented as paper AIAA-2006-6173 to AIAA/AAS Astrodynamics Specialist Conference, 21-24 August 2006, USA. See also http://arxiv.org/ftp/physics/papers/0701/0701073.pdf

Introduction

Brief information about space particles and space environment. In Earth's atmosphere at altitudes between 200 - 400 km, the concentration of ions reaches several million per cubic cm. In the interplanetary medium at Earth orbit, the concentration of protons from the Solar Wind reaches 3 - 70 particles per cubic cm. In an interstellar medium the average concentration of protons is about one particle in 1 cm^3, but in the space zones HII (planetary nebulas), which occupy about 5% of interstellar space, the average particle density may be 10^{-20} g/cm^3 (10^6 particles in 1 см^3). If we can collect these space particles from a large area, accelerate and brake them, we can get the high speed and braking of space apparatus and to generate energy. The author is suggesting the method of collection and implementations of it for propulsion and braking systems and electric generators. He developed the initial theory of these systems.

Short Description of the Implémentation

A *Primary Ramjet* propulsion engine is shown in Figure 1-1. Such an engine can work in one charge environment. For example, the surrounding region of space medium contains the positive charge particles (protons, ions). The engine has two plates 1, 2, and a source of electric voltage and energy (storage) 3. The plates are made from a thin dielectric film covered by a conducting layer. As the plates may be a net. The source can create an electric voltage U and electric field (electric intensity E) between the plates. One also can collect the electric energy from plate as an accumulator.

The engine works in the following way. Apparatus are moving (in left direction) with velocity V (or particles 4 are moving in right direction). If voltage U is applied to the plates, it is well-known that main electric field is only between plates. If the particles are charged positive (protons, positive ions) and the first and second plate are charged positive and negative, respectively, then the particles are accelerated between the plates and achieve the additional velocity $v > 0$. The total velocity will be $V+v$ behind the engine (Figure 1a). This means that the apparatus will have thrust $T > 0$ and spend electric energy $W < 0$ (bias, displacement current). If the voltage $U = 0$, then $v = 0$, $T = 0$, and $W = 0$ (Figure 1-1b).

If the first and second plates are charged negative and positive, respectively, the voltage changes sign Assume the velocity v is satisfying $-V < v < 0$. Thus the particles will be broken and the engine (apparatus) will have drag and will also be broken. The engine transfers broke vehicle energy into electric (bias, displacement) current. That energy can be collected and used. Note that velocity v cannot equal $-V$. If v were equal to $-V$, that would mean that the apparatus collected positive particles, accumulated a big positive charge and then repelled the positive charged particles.

If the voltage is enough high, the brake is the highest (Figure 1-1d). Maximum braking is achieved when $v = -2V$ ($T < 0$, $W = 0$). Note, the v cannot be more then $-2V$, because it is full reflected speed.

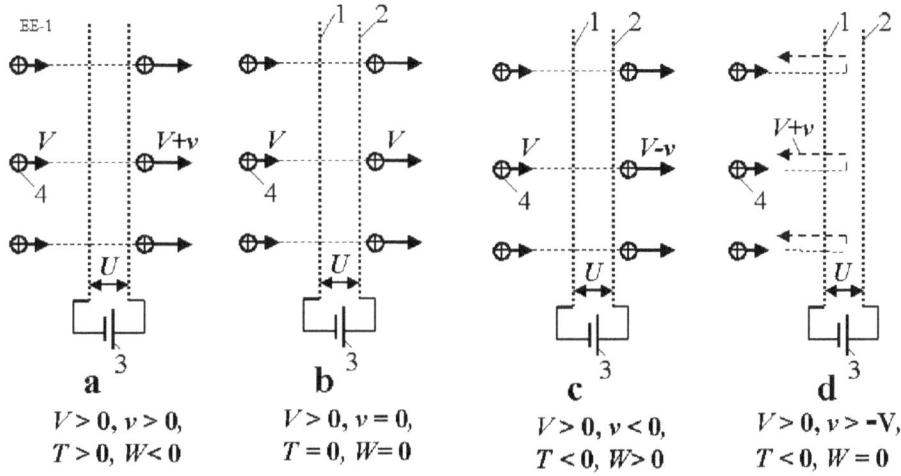

Figure 1-1. Explanation of primary Space Ramjet propulsion (engine) and electric generator (in braking), a) Work in regime thrust; b) Idle; c) Work in regime brake. d) Work in regime strong brake (full reflection). Notation: 1, 2 - plate (film, thin net) of engine; 3 - source of electric energy (voltage U); 4 - charged particles (protons, ions); V - speed of apparatus or particles before engine (solar wind); v - additional speed of particles into engine plates; T - thrust of engine; W - energy (if $W < 0$ we spend energy).

AB-Ramjet engine. The suggested Ramjet is different from the primary ramjet. The suggested ramjet has specific electrostatic collector 5 (Figure 1-2a,c,d,e,f,g). Other authors said the idea of space matter collection.

But they did not give the principal design of collector. Their electrostatic collector cannot work. Really, for charging of collector we must move away from apparatus the charges. The charged collector attracts the same amount of the charged particles (charged protons, ions, electrons) from space medium. They discharged collector. All your work will be idle. That cannot work.

The electrostatic collector cannot absorb a matter (as offered some inventors) because it can absorb ONLY opposed charges particles, which will be discharged the initial charge of collector. Physic law of conservation of charges does not allow changing the charges of particles.

The suggested collector and ramjet engine have a special design (thin film, net, special form of charge collector, particle accelerator). The collector/engine passes the charged particles ACROSS (through) the installation and changes their energy (speed), deflecting and focusing them. That is why we refer to this engine as the *AB-Ramjet engine*. It can create thrust or drag, extract energy from the kinetic energy of particles or convert the apparatus' kinetic energy into electric energy, and deflect and focus the particle beam. The collector creates a local environment in space because it deletes (repeals) the same charged particles (electrons) from apparatus and allows the Ramjet to work when the apparatus speed is close to zero. The author developed the theory of the electrostatic collector. The conventional electric engine cannot work in usual plasma without the main part of the AB-engine - the special pervious electrostatic collector.

The plates of the suggested engine are different from the primary engine. They have a concentrically septa (partitions) which create additional radial electric fields (electric intensity) (Figure 1-2b). They straighten, deflect and focus the particle beams and improve the efficiency coefficient of the engine.

The central charge can have a different form (core) and design (Figure2 c,d,e,f,g,h). It may be:

(1) a sphere (Figure 1-2c) having a thin cover of plastic film and a very thin (some nanometers) conducting layer (aluminum), with the concentrically spheres inserted one into the other (Figure 1-2d),
(2) a net formed from thin wires (Figure 1-2e);
(3) a cylinder (without butt-end)(Figure 1-2f); or
(4) a plate (Figure 1-2g).

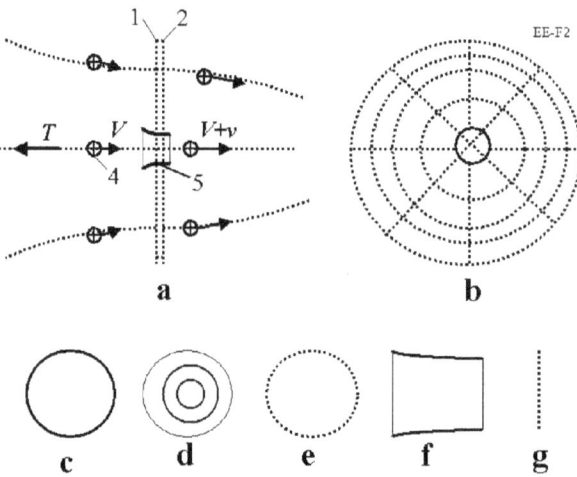

Figure 1-2. Space AB-Ramjet engine with electrostatic collector (core). a) Side view; b) Front view; c) Spherical electrostatic collector (ball); d) Concentric collector; e) cellular (net) collector; f) cylindrical collector without cover butt-ends; g) plate collector (film or net).

The design is chosen to produce minimum energy loss (maximum particle transparency). The safety (from discharging, emission of electrons) electric intensity in a vacuum is 10^8 V/m for an outer conducting layer and negative charge. The electric intensity is more for an inside conducting layer and thousands of times more for positive charge.

The engine plates are attracted one to the other. They can have different designs (Figure 1-3a - 3d). In the rotating film or net design (Figure 1-3a), the centrifugal force prevents contact between the plates. In the inflatable design (Figure 1-3b), the low pressure gas prevents plate contact. A third design has (inflatable) rods supporting the film or net (Figure 1-3c). The fourth design is an inflatable toroid which supports the distance between plates or nets (Figure 1-3d).

Electric gun. The simplest electric gun (linear particle accelerator) for charging an apparatus ball is presented in Figure 1-4. The design is a long tube (up 10 m) which creates a strong electric field along the tube axis (100 MV/m and more). The gun consists of the tube with electrical isolated cylindrical electrodes, ion source, microwave frequency energy source, and voltage multiplier. This electric gun can accelerate charged particles up 1000 MeV. Electrostatic lens and special conditions allow the creation of a focusing and self-focusing beam which can transfer the charge and energy long distances into space. The engine can be charged from a satellite, a space ship, the Moon, or a top atmosphere station. The beam may also be used as a particle beam weapon.

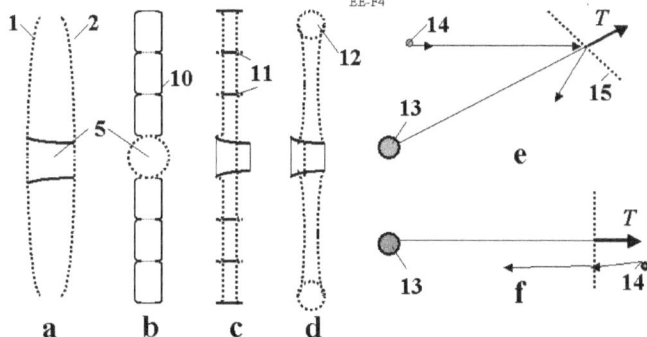

Figure 1-3. Possible design of the main part of ramjet engine. a) Rotating engine; b) Inflatable engine (filled by gas); c) Rod engine; d) Toroidal shell engine, e) AB-Ramjet engine in brake regime, f) AB-Ramjet engine in thrust regime. Notation: 10 - film shells (fibers) for support thin film and creating a radial electric field; 11 - Rods for a support the film or net; 12 - inflatable toroid for support engine plates; 13 - space apparatus; 14 - particles; 15 - AB-Ramjet.

Approximately tens years ago, the conventional linear pipe accelerated protons up to 40 MeV with a beam divergence of 10^{-3} radian. However, acceleration of the multi-charged heavy ions may result in significantly more energy.

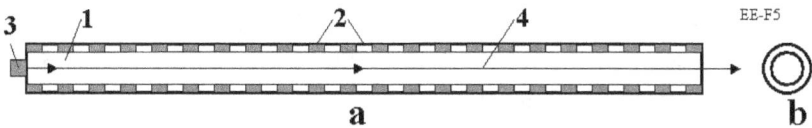

Figure 1-4. Electric gun for charging AB-Ramjet engine and transfer charges (energy) in long distance. a) Side view, b) Front view. *Notations:* 1 - gun tube, 2 - opposed charged electrodes, 3 - source of charged particles (ions, electrons), 4 - particles beam.

At present, the energy gradients as steep as 200 GeV/m have been achieved over millimeter-scale distances using laser pulsars. Gradients approaching 1 GeV/m are being produced on the multi-centimeter-scale with electron-beam systems, in contrast to a limit of about 0.1 GeV/m for radio-frequency acceleration alone. Existing electron accelerators such as SLAC <http://en.wikipedia.org/wiki/SLAC> could use electron-beam afterburners to increase the intensity of their particle beams. Electron systems in general can provide tightly collimated, reliable beams while laser systems may offer more power and compactness.

Conclusion

The primary research and computations of the suggested AB-engine show the numerous possibilities and perspectives of the space AB-ramjet engines. The density of the charged space particles is very small. But the proposed electrostatic collector can effectively gather the particles from a huge surrounding area and accelerate or brake them, generating thrust or braking on the order of several Newtons. The high speed solar wind allows simultaneously obtainment of useful drag (thrust) and great electrical energy. The simplest electrostatic gatherer accelerates a 100 kg probe up to a velocity of 100 km/s. The probe offers flights into Mars orbit of about 70 days, to Jupiter orbit in about 150 days, to Saturn orbit in about 250 days, to Uranus orbit in about 450 days, to Neptune orbit in about 650 days, and to Pluto orbit in about 850 days.

The suggested electric gun is simple and can transfer energy (charge by electron beam) over a long distance to other space apparatus.

The author has developed the initial theory and the initial computations to show the possibility of the offered concepts. He calls on scientists, engineers, space organizations, and companies to research and develop the proposed perspective concepts.

2. Beam Space Propulsion*

The author offers a revolutionary method - non-rocket transfer of energy and thrust into Space with distance of millions kilometers. The author has developed theory and made the computations. The method is more efficient than transmission of energy by high-frequency waves. The method may be used for space launch and for acceleration the spaceship and probes for very high speeds, up to relativistic speed by current technology. Research also contains prospective projects which illustrate the possibilities of the suggested method.

Description of Innovation

Innovative installation for transfer energy and impulse includes (Figure 2-1): the ultra-cold plasma injector, electrostatic collector, electrostatic electro-generator-thruster-reflector, and space apparatus. The plasma injector creates and accelerates the ultra-cold low density plasma.

The Installation works the following way: the injector-accelerator forms and injects the cold neutral plasma beam with high speed in spaceship direction. When the beam reaches the ship, the electrostatic collector of spaceship collects and separates the beam ions from large area and passes them through the engine-electric

---

* Presented as paper AIAA-2006-7492 to Conference "Space-2006", 19-21 September, 2006, San-Jose, CA, USA. See also http://arxiv.org/ftp/physics/papers/0701/0701057.pdf

generator or reflects them by electrostatic mirror. If we want to receive the thrust in the near beam direction (± 90°) and electric energy, the engine works as thruster (accelerator of spaceship and braker of beam) in beam direction and electric generator. If we want to get thrust in opposed beam direction, the space engine must accelerate the beam ions and spend energy. If we want to have maximum thrust in beam direction, the engine works as full electrostatic mirror and produces double thrust in the beam direction (full reflection of beam back to injector). The engine does not spend energy for full reflection.

The thrust is controlled by the electric voltage between engine nets, the thrust direction is controlled by the engine nets angle to beam direction. Note, the trust can brake the ship (decrease the tangential ship speed) and far ship (located out of Earth orbit) can return to the Earth by Sun gravity.

Note also, the Earth atmosphere absorbs and scatters the plasma beam and the beam injector must be located on Earth space mast or tower (up 40 ÷ 60 km) or the Moon. Only high energy beam can break through atmosphere with small divergence. The advantage: the injector has a reflector and when the ship locates not far from the injector the beam will be reflected a lot of times and thrust increases in thousand times at start
The proposed engine may be also used as AB-ramjet engine, utilizing the Solar wind or interstellar particles.

Figure 2-1. Long distance space transfer of electric energy, matter, and momentum (thrust). Notation are: 1 - injector-accelerator of neutral ultra-cold plasma (ions and electrons), 2 - plasma beam, 3 - space ship or planetary team, 4 - electrostatic ions collector (or magnetic collector), 5 - braking electric nets (electrostatic electro-generator-thruster-reflector), 6 - thrust.

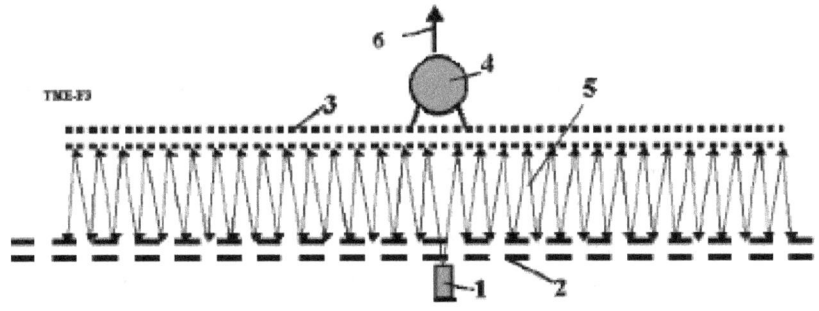

Figure 2-2. Multi-reflection start of the spaceship having proposed engine. Notation are: 1 - injector-accelerator of cold ions or plasma, 2 , 3 - electrostatic reflectors, 4 - space ship, 5 - plasma beam, 6 - the thrust.

The electrostatic collector and electrostatic generator-thruster-reflector proposed and described in chapter 1 above. The main parts are presented below.

A *Primary Ramjet* propulsion engine is shown in Figure 2-1 Chapter 1. Such an engine can work in charged environment. For example, the surrounding region of space medium contains positive charge particles (protons, ions). The engine has two plates 1, 2, and a source of electric voltage and energy (storage) 3. The plates are made from a thin dielectric film covered by a conducting layer. The plates may be a net. The source can create an electric voltage U and electric field (electric intensity E) between the plates. One also can collect the electric energy from plate as an accumulator.

The engine works in the following way. Apparatus are moving (in left direction) with velocity V (or particles 4 are moving in right direction). If voltage U is applied to the plates, it is well-known that main electric field is only between plates. If the particles are charged positive (protons, positive ions) and the first and second plate are charged positive and negative, respectively, then the particles are accelerated between the plates and achieve the additional velocity $v > 0$. The total velocity will be $V+v$ behind the engine (Figure 2-1a, Ch. 2). This means that the apparatus will have thrust $T > 0$ and spend electric energy $W < 0$ (bias, displacement current). If the voltage $U = 0$, then $v = 0$, $T = 0$, and $W = 0$ (Figure 1-1b, Ch.1).

If the first and second plates are charged negative and positive, respectively, the voltage changes sign. Assume the velocity v is satisfying $-V < v < 0$. Thus the particles will be braked and the engine (apparatus) will have drag and will also be braked. The engine transfers braked vehicle energy into electric (bias, displacement) current. That energy can be collected and used. Note that velocity v cannot equal $-V$. If v were equal to $-V$, that would mean that the apparatus collected positive particles, accumulated a big positive charge and then repelled the positive charged particles.

If the voltage is high enough, the brake is the highest (Figure1-1d, Ch.1). Maximum braking is achieved when $v = -2V$ ($T < 0$, $W = 0$). Note, the v cannot be more then $-2V$, because it is full reflected speed.

AB-Ramjet engine. The suggested Ramjet is different from the primary ramjet. The suggested ramjet has specific electrostatic collector 5 (Figure 1-2a,c,d,e,f,g, Ch. 1). Other authors have outline the idea of space matter collection, but they did not describe and not research the principal design of collector. Really, for charging of collector we must move away from apparatus the charges. The charged collector attracts the same amount of the charged particles (charged protons, ions, electrons) from space medium. They discharged collector, work will be idle. That cannot be useful.

The electrostatic collector cannot adsorb matter (as offered some inventors) because it can adsorb ONLY opposed charges particles, which will be discharged the initial charge of collector. Physic law of conservation of charges does not allow to change charges of particles.

The suggested collector and ramjet engine have a special design (thin film, net, special form of charge collector, particle accelerator). The collector/engine passes the charged particles ACROSS (through) the installation and changes their energy (speed), deflecting and focusing them. That is why we refer to this engine as the *AB-Ramjet engine*. It can create thrust or drag, extract energy from the kinetic energy of particles or convert the apparatus' kinetic energy into electric energy, and deflect and focus the particle beam. The collector creates a local environment in space because it deletes (repeals) the same charged particles (electrons) from apparatus and allows the Ramjet to work when the apparatus speed is close to zero. The author developed the theory of the electrostatic collector and published it in [26]. The conventional electric engine cannot work in usual plasma without the main part of the AB-engine - the special pervious electrostatic collector.

The plates of the suggested engine are different from the primary engine. They have concentric partitions which create additional radial electric fields (electric intensity) (Figure 1-2b, Ch.1). They straighten, deflect and focus the particle beams and improve the efficiency coefficient of the engine.

The central charge can have a different form (core) and design (Figure 2-2 c,d,e,f,g,h, Ch.2). It may be:

(1) a sphere (Figure 1-2c, Ch. 1) having a thin cover of plastic film and a very thin (some nanometers) conducting layer (aluminum), with the concentric spheres inserted one into the other (Figure 1-2d, Ch. 1),
(2) a net formed from thin wires (Figure 1-2e, Ch. 1);
(3) a cylinder (without butt-end)(Figure 1-2f, Ch. 1); or
(4) a plate (Figure 1-2g, Ch. 1).

The design is chosen to produce minimum energy loss (maximum particle transparency - see section "Theory" in [2] Ch.3). The safety (from discharging, emission of electrons) electric intensity in a vacuum is 10^8 V/m for an outer conducting layer and negative charge. The electric intensity is more for an inside conducting layer and thousands of times more for positive charge.

The engine plates are attracted one to the other (see theoretical section in [2] Ch.3). They can have various designs (Figure 1-3a - 3d, Ch. 1). In the rotating film or net design (Figure 1-3a, Ch. 1), the centrifugal force prevents contact between the plates. In the inflatable design (Figure 1-3b, Ch. 1), the low pressure gas prevents plate contact. A third design has (inflatable) rods supporting the film or net (Figure 1-3c, Ch. 1). The fourth design is an inflatable toroid which supports the distance between plates or nets (Figure 1-3d, Ch.1).

Note, the AB-ramjet engine can work using the neutral plasma. The ions will be accelerated or braked, the electrons will be conversely braked or accelerated. But the mass of the electrons is less then the mass of ions in thousands times and AB-engine will produce same thrust or drag.

Plasma accelerator. The simplest linear plasma accelerator (principle scheme of linear particle accelerator) for plasma beam is presented in Figure 1-4, Ch. 1. The design is a long tube (up 10 m) which creates a strong electric field along the tube axis (100 MV/m and more). The accelerator consists of the tube with electrical isolated cylindrical electrodes, ion source, and voltage multiplier. The accelerator increases speed of ions, but in end of tube into ion beam the electrons are injected. This plasma accelerator can accelerate charged particles up 1000 MeV. Electrostatic lens and special conditions allow the creation of a focusing and self-focusing beam which can transfer the charge and energy long distances into space. The engine can be charged from a satellite, a spaceship, the Moon, or a top atmosphere station. The beam may also be used as a particle beam weapon.

Approximately ten years ago, the conventional linear pipe accelerated protons up to 40 MeV with a beam divergence of 10^{-3} radian. However, acceleration of the multi-charged heavy ions may result in significantly more energy.

At present, the energy gradients as steep as 200 GeV/m have been achieved over millimeter-scale distances using laser pulsers. Gradients approaching 1 GeV/m are being produced on the multi-centimeter-scale with electron-beam systems, in contrast to a limit of about 0.1 GeV/m for radio-frequency acceleration alone. Existing electron accelerators such as SLAC <http://en.wikipedia.org/wiki/SLAC> could use electron-beam afterburners to increase the intensity of their particle beams. Electron systems in general can provide tightly collimated, reliable beams while laser systems may offer more power and compactness.

The cool plasma beam carries three types of energy: kinetic energy of particles, ionization, and dissociation energy of ions and molecules. That carries also particle mass and momentum. The AB-Ramjet engine (described over) can utilize only kinetic energy of plasma particles and momentum. The particles are broken and produce an electric current and thrust or reflected and produce only thrust in the beam direction. If we want to collect a plasma matter and to utilize also the ionization energy of plasma (or space environment) ions and dissociation energy of plasma molecules we must use the modified AB-Ramjet engine described below (Figure 2-3).

The modified AB-engine has magnetic collector (option), three nets (two last nets may be films), and issue voltage (that also may be an electric load). The voltage, U, must be enough for full braking of charged particles. The first two nets brake the electrons and precipitate (collect) the electrons on the film 2 (Figure 2-3). The last couple of film (2, 3 in Figure 2-3) brakes and collects the ions. The first couple of nets accelerate the ions that are way the voltage between them must be double.

The collected ions and electrons have the ionized and dissociation energy. This energy is significantly (up 20 - 150 times) more powerful then chemical energy of rocket fuel but significantly less then kinetic energy of particles (ions) equal U (in eV) (U may be millions volts). But that may be used by ship. The ionization energy conventionally pick out in photons (light, radiation) which easy are converted in a heat (in closed vessel), the dissociation energy conventionally pick out in heat.

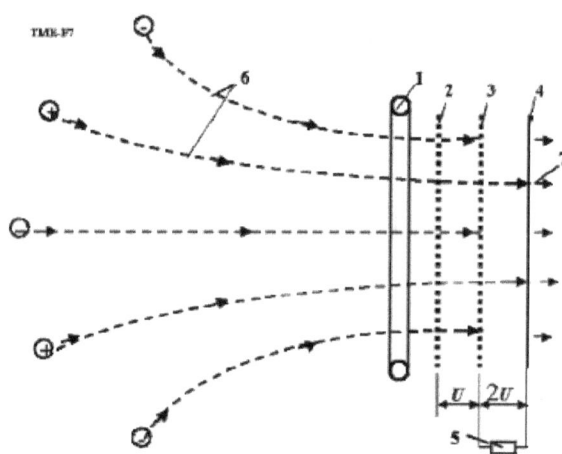

Figure 2-3. AB-engine which collected matter of plasma beam, kinetic energy of particles, energy ionization and dissociation. Notations: 1 - magnetic collector; 2 - 4 - plates (films, nets) of engine; 5 - electric load; 6 - particles of plasma; 7 - radiation. U - voltage between plates (nets).

The light energy may be used in the photon engine as thrust (Figure 2-4a) or in a new power laser (Figure 2-4b). The heat energy may be utilized conventional way (Figure 2-4c). The offered new power laser (Figure 4b) works the following way. The ultra-cool rare plasma with short period of life time located into cylinder. If we press it (decrease density of plasma) the electrons and ions will connect and produce photons of very closed energy (laser beam). If we compress very quickly by explosion the power of beam will be high. The power is only limited amount of plasma energy.

After recombination ions and electrons we receive the conventional matter. This matter may be used as nuclear fuel (in thermonuclear reactor), medicine, food, drink, oxidizer for breathing, etc.

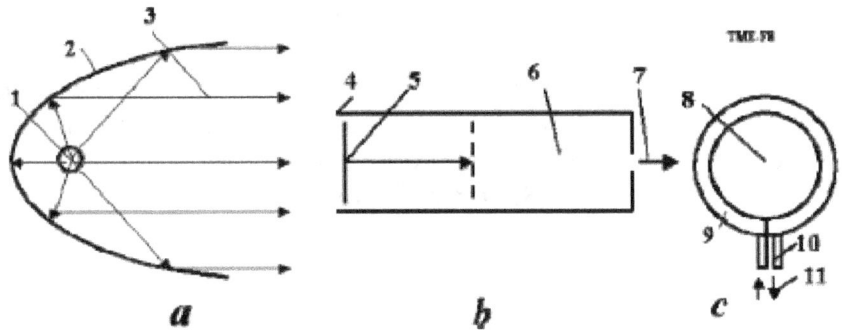

Figure 2-4. Conversion of ionization energy into radiation and heat. *a* - photon engine; *b* - power laser (light beamer); *c* - heater. Notations: 1 - recombination reactor; 2 - mirror; 3 - radiation (light) beam; 5 - piston; 6 - volume filled by cold rare plasma; 7 - beam; 8 - plasma; 9 - heat exchanger; 10 - enter and exit of hear carrier; 11 - heat carrier.

Conclusion

The offered idea and method use the AB-Ramjet engine suggested by author in 1982 [2- 4] and detail developed in [2]. The installation contains an electrostatic particle collector suggested in 1982 and detail developed in [2]. The propulsion-reflected system is light net from thin wire, which can have a large area (tens km) and allows to control thrust and thrust direction without turning of net (Figure1). This new method uses the ultra-cold full neutral relativistic plasma and having small divergence. The method may be used for acceleration space apparatus (up relativistic speed) for launch and landing Space apparatus to small planets (asteroids, satellites) without atmosphere. For Earth offered method will be efficiency if we built the tower (mast) about 40 ÷ 80 km height. At present time that is the most realistic method for relativistic probe.

3. Space Magnetic Sail*

The first reports on the "Space Magnetic Sail" concept appeared more 30 years ago. During the period since some hundreds of research and scientific works have been published, including hundreds of research report by professors at major famous universities. The author herein shows that all these works related to Space Magnetic Sail concept are technically incorrect because their authors did not take into consideration that solar wind impinging a MagSail magnetic field creates a particle magnetic field opposed to the MagSail field. In the incorrect works, the particle magnetic field is hundreds times stronger than a MagSail magnetic field. That means all the laborious and costly computations in revealed in such technology discussions are useless: the impractical findings on sail thrust (drag), time of flight within the Solar System and speed of interstellar trips are essentially worthless working data! The author reveals the correct equations for any estimated performance of a Magnetic Sail as well as a new type of Magnetic Sail (without a matter ring).

New Electrostatique MagSail (EMS)

The conventional MagSail with super-conductive ring has big drawbacks:

---

* Presented as paper AIAA-2006-8148 to 14th AIAA/AHI Space Planes and Hypersonic Systems and Technologies Conference, 6 - 9 Nov 2006 National Convention Centre, Canberra, Australia. See also [2, Ch.4],
http://www.scribd.com/doc/24057071 .
http://www.archive.org/details/NewConceptsIfeasAndInnovationsInAerospaceTechnologyAndHumanSciences

1. It is very difficult to locate gigantic (tens of km radius) ring in outer space.
2. It is difficult to insert a big energy into superconductive ring.
3. Super-conductive ring needs a low temperature to function at all. The Sun heats all bodies in the Solar System to a temperature higher then temperature of super-conductive materials.
4. The super-conductive ring explodes if temperature is decreased over critical value.
5. It is difficult to control the value of MagSail thrust and the thrust direction.

The author offers new Electrostatic MagSail (EMS). The innovation includes the central positive charged small ball and a negative electronic equal density ring rotated around the ball (Figure 3-1).

The suggested EMS has the following significant advantages in comparison with conventional MagSail:

(1) No heavy super-conductive large ring.
(2) No cooling system for ring is required.
(3) Electronic ring is safe.
(4) The thrust (ring radius) easy changes by changing of ball charge.

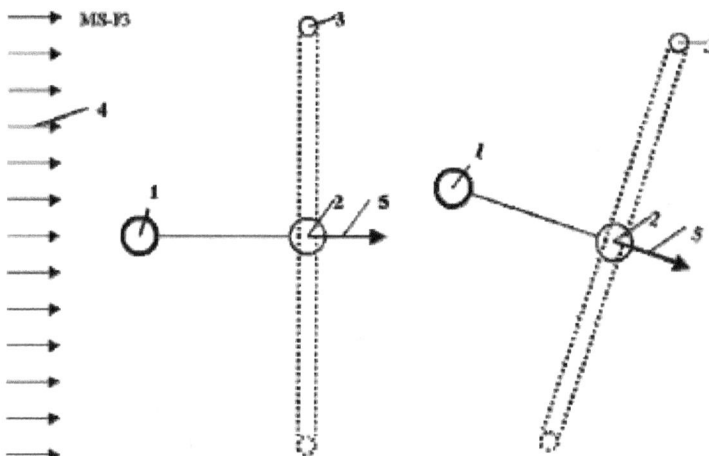

Figure 3-1. Electrostatic MagSail. Notations: 1-Spaceship; 2-Positive charged ball; 3–electrical ring; 4-solar wind; 5-EMS drag. In right side the EMS in turn position.

4. High Speed AB-Solar Sail*

The Solar sail is a large thin film used to collect solar light pressure for moving of space apparatus. Unfortunately, the solar radiation pressure is very small about 9 μN/mm at Earth's orbit. However, the light force significantly increases up to 0.2 - 0.35 N/mm near the Sun. The author offers his research on a new revolutionary highly reflective solar sail which flyby (after special maneuver) near Sun and attains velocity up to 400 km/sec and reaching far planets of the Solar system in short time or enable flights out of Solar system. New, highly reflective sail-mirror allows

---

* This work is presented as paper AIAA-2006-4806 for 42 Joint Propulsion Conference, Sacramento, USA, 9-12 July, 2006. See also http://arxiv.org/ftp/physics/papers/0701/0701073.pdf or [2, Ch.5].

avoiding the strong heating of the solar sail. It may be useful for probes close to the Sun and Mercury and Venus.

Description and Innovations

Description. The suggested AB space sail is presented in Figure 4-1. It consists of: a thin high reflection film (solar sail) supported by an inflatable ring (or other method), space apparatus connected to solar sail, a heat screen defends the apparatus from solar radiation.

The thin film includes millions of very small prisms (angle 45°, side 3 ÷ 30 μm). The solar light is totally reflected back into the incident medium. This effect is called total internal reflection. Total internal reflection is used in the proposed reflector. As it is shown in [1, Ch.12] the light absorption is very small ($10^{-5} \div 10^{-7}$) and radiation heating is small (see computation section).

Another possible design for the suggested solar sail is presented in Figure 4-2. Here solar sail has concave form (or that plate is made like Fresnel mirror).

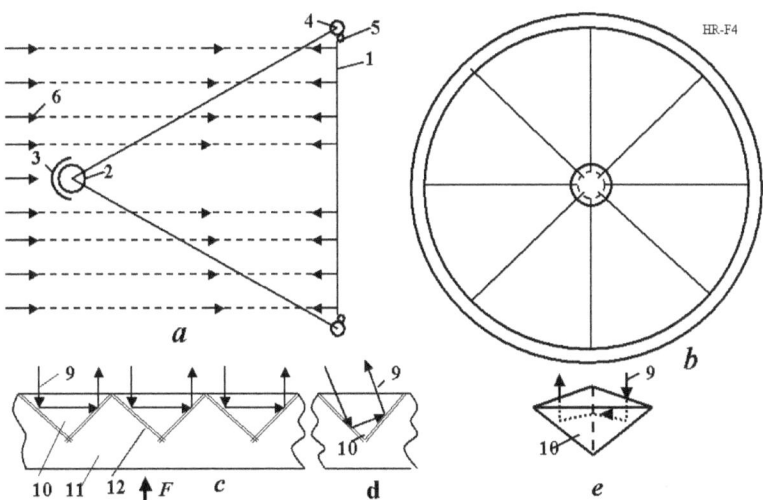

Figure 4-1. High reflective space AB-sail. (a) Side view of AB-sail; (b) Front view; (c) cross-section of sail surface; (d) case of non-perpendicular solar beam; (e) triangle reflective sell. Notation: 1 - thin film high reflective AB-mirror, 2 - space apparatus, 3 - high reflective heat screen (shield) of space apparatus, 4- inflatable support thin film ring, 5 - inflatable strain ring, 6 - solar light, 9 - solar beam, 10 - reflective sell, 11 - substrate, 12 - gap.

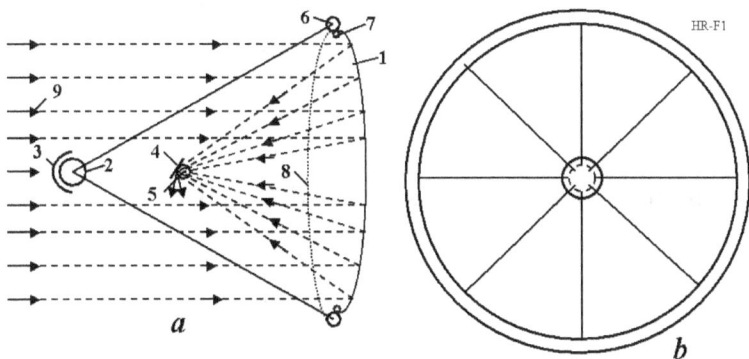

Figure 4-2. AB highly reflective solar sail with concentrator. (a) side view; (b) front view. Notation: 1 - highly reflective AB mirror (it may have a Fresnel form), 2 - space apparatus, 3 - high reflective heat screen, 4 - control mirror, 5 - reflected solar beam, 6 - inflatable support thin film ring, 7 - inflatable strain ring, 8 - thin transparent film, 9 - solar beam.

The sail concentrates solar light on a small control mirror 4. That mirror allows re-directed (reflected) solar beam and to change value and direction of the sail thrust without turning the large solar sail. Between thin films 1, 8 there is a small gas pressure which supports the concave form of reflector 1. Concentration of energy can reach $10^3 \div 10^4$ times, temperature greater than 5000 °K. This energy may be very large. For the sail of 200×200 m, at Earth orbit the energy is 5.6×10^4 kW. This energy may be used for apparatus propulsion or other possibilities (see [5]), for example, to generate electricity. The concave reflector may be also utilized for long-distance radio communication.

The trajectory of the high speed solar AB-sail is shown in Figure 4-3. The sail starts from Earth orbit. Then is accelerated by a solar light to up 11 km/s in opposed direction of Earth moving around Sun and leaves Earth gravitational field. The Earth has a speed about 29 km/s in its around Sun orbit. The sail will be have 29 -11=18 km/s. That is braked and moves to Sun (trajectory 4). Near the Sun the reflector is turned for acceleration to get a high speed (up to 400 km/s) from a powerful solar radiation. The second solar space speed is about 619 km/s. If AB sail makes three small revolutions around Sun, it can then reach speed of a 1000 km/s and leaves the Solar system with a speed about 400 km/s. Suggested highly reflective screen protects the apparatus from an excessive solar heating. Note, the offered AB sail allows also to brake an apparatus very efficiency from high speed to low speed. If we send AB sail to another star, it can brake at that star and became a satellite of the star (or a planet of that solar system).

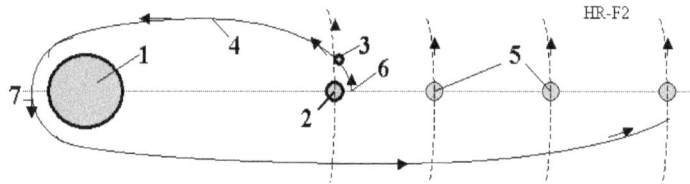

Figure 4-3. Maneuvers of AB solar sail for reaching a high speed: braking for flyby near Sun, great acceleration from strong solar radiation and flight away to far planets or out of our Solar system. Notation: 1 - Sun, 2 - Earth, 3 AB Solar sail, 4 - trajectory of solar sail to Sun, 5 - other planets, 6, 7 - speed of solar sail.

Conclusion

The suggested new AB sail can fly very close to the Sun's surface and get high speed which is enough for quick flight to far planets and out of our Solar System. Advantages allow: 1) to get very high speed up 400 km/s; 2) easy to control an amount and direction of thrust without turning a gigantic sail; 3) to utilize of the solar sail as a power generator (for example, electricity generator); 4) to use the solar sail for long-distance communication systems.

The same researches were made by author for solar wind sail and other propulsion [1]-[4].

5. Transfer of Electricity in Outer Space[*]

---

[*] Presented as Bolonkin's paper AIAA-2007-0590 to 45th AIAA Aerospace Science Meeting, 8 - 11 January 2007, Reno, Nevada, USA. See also http://arxiv.org/ftp/physics/papers/0701/0701058.pdf .

Author offers conclusions from his research of a revolutionary new idea - transferring electric energy in the hard vacuum of outer space wirelessly, using a plasma power cord as an electric cable (wire). He shows that a certain minimal electric currency creates a compressed force that supports the plasma cable in the compacted form. A large energy can be transferred hundreds of millions of kilometers by this method. The required mass of the plasma cable is only hundreds of grams. He computed the macroprojects: transference of hundreds kilowatts of energy to Earth's Space Station, transferring energy to the Moon or back, transferring energy to a spaceship at distance 100 million of kilometers, the transfer energy to Mars when one is located at opposed side of the distant Sun, transfer colossal energy from one of Earth's continents to another continent (for example, between Europe-USA) wirelessly—using Earth's ionosphere as cable, using Earth as gigantic storage of electric energy, using the plasma ring as huge MagSail for moving of spaceships. He also demonstrates that electric currency in a plasma cord can accelerate or brake spacecraft and space apparatus.

Innovations and Bref Descriptions

The author offers the series of innovations that may solve the many macro-problems of transportation energy in space, and the transportation and storage energy within Earth's biosphere. Below are some of them.

(1) Transfer of electrical energy in outer space using the conductive cord from plasma. Author solved the main problem - how to keep plasma cord in compressed form. He developed theory of space electric transference, made computations that show the possibility of realization for these ideas with existing technology. The electric energy may be transferred in hundreds millions of kilometers in space (include Moon and Mars).
(2) Method of construction for space electric lines and electric devices.
(3) Method of utilization of the plasma cable electric energy.
(4) A new very perspective gigantic plasma MagSail for use in outer space as well as a new method for connection the plasma MagSail to spaceship.
(5) A new method of projecting a big electric energy through the Earth's ionosphere.
(6) A new method for storage of a big electric energy used Earth as a gigantic spherical condenser.
(7) A new propulsion system used longitudinal (cable axis) force of electric currency.

Below there are some succinct descriptions of some constructions made possible by these revolutionary ideas.

1. Transferring Electric Energy in Space

The electric source (generator, station) is connected to a space apparatus, space station or other planet by two artificial rare plasma cables (Figure 5-1a). These cables can be created by plasma beam [2] sent from the space station or other apparatus.

The plasma beam may be also made the space apparatus from an ultra-cold plasma [2] when apparatus starting from the source or a special rocket. The plasma cable is self-supported in cable form by magnetic field created by electric currency in plasma cable because the magnetic field produces a magnetic pressure opposed to a gas dynamic plasma pressure (teta-pinch)(Figure 5-2). The plasma has a good conductivity (equal silver and more) and the plasma cable can have a very big cross-section area (up thousands of square meter). The plasma conductivity does not depend on its density. That way the plasma cable has a no large resistance although the length of plasma cable is hundreds millions of kilometers. The needed minimum electric currency from parameters of a plasma cable researched in theoretical section of this article.

The parallel cables having opposed currency repels one from other (Figure 1a). They also can be separated by a special plasma reflector as it shown in figs. 5-1b, 5-1c. The electric cable of the plasma transfer can be made circular (Figure 5-1c).

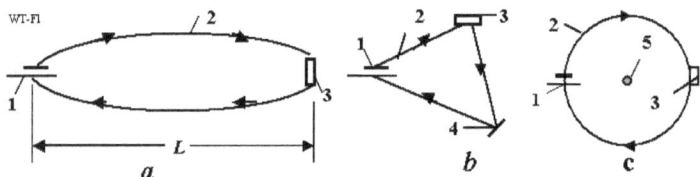

Figure 5-1. Long distance plasma transfer electric energy in outer space. a - Parallel plasma transfer, b - Triangle plasma transfer, c - circle plasma transfer. Notations: 1 - current source (generator), 2 - plasma wire (cable), 3 - spaceship, orbital station or other energy addresses, 4 - plasma reflector, 5 - central body.

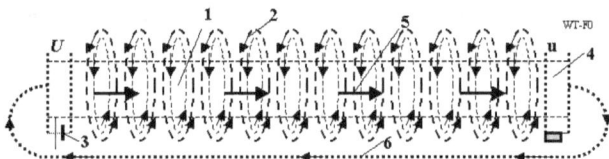

Figure 5-2. A plasma cable supported by self-magnetic field. Notations: 1 -plasma cable, 2 - compressing magnetic field, 3 - electric source, 4 - electric receiver, 5 - electric currency, 6 - back plasma line.

The radial magnetic force from a circle currency may be balanced electric charges of circle and control body or/and magnetic field of the space ship or central body (see theoretical section). The circle form is comfortable for building the big plasma cable lines for spaceship not having equipment for building own electric lines or before a space launch. We build small circle and gradually increase the diameter up to requisite value (or up spaceship). The spaceship connects to line in suitable point. Change the diameter and direction of plasma circle we support the energy of space apparatus. At any time the spaceship can disconnect from line and circle line can exist without user.

The electric tension (voltage) in a plasma cable is made two nets in issue electric station (electric generator). The author offers two methods for extraction of energy from the electric cable (Figure3) by customer (energy addresses). The plasma cable currency has two flows: electrons (negative) flow and opposed ions (positive) flow in one cable. These flows create an electric current. (It may be instances when ion flow is stopped and current is transferred only the electron flow as in a solid metal or by the ions flow as in a liquid electrolyte. It may be the case when electron-ion flow is moved in same direction but electrons and ions have different speeds). In the first method the two nets create the opposed electrostatic field in plasma cable (resistance in the electric cable) (figs. 5-1, 5-3b). This apparatus resistance utilizes the electric energy for the spaceship or space station. In the second method the charged particles are collected a set of thin films (Figure 3a) and emit (after utilization in apparatus) back into continued plasma cable (Figure 5-3a).

Figure 5-3c presents the plasma beam reflector. That has three charged nets. The first and second nets reflect (for example) positive particles, the second and third nets reflected the particles having an opposed charge.

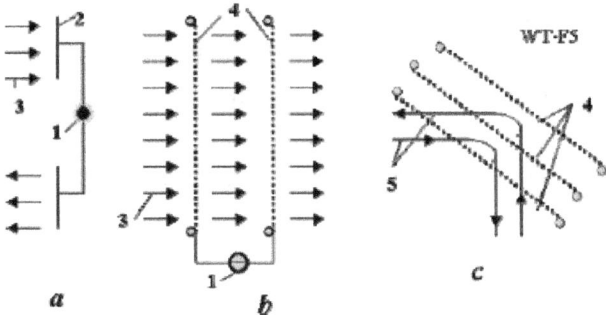

Figure 5-3. Getting the plasma currency energy from plasma cable. a - getting by two thin conducting films; b - getting two nets which brake the electric current flux; c - plasma reflector. Notations: 1 - spaceship or space station, 2 - set (films) for collect (emit) the charged particles, 3 - plasma cable, 4 - electrostatic nets.

2. Transmitting of the Electric Energy to Satellite, Earth's Space Station, or Moon.

The suggested method can be applied for transferring of electric energy to space satellites and the Moon. For transmitting energy from Earth we need a space tower of height up 100 km, because the Earth's atmosphere will wash out the plasma cable or we must spend a lot of energy for plasma support. The design of solid, inflatable, and kinetic space towers are revealed in [1]-[13]-[4] and in Review of Space Towers into this Collction.

It is possible this problem may be solved with an air balloon located at 30-45 km altitude and connected by conventional wire with Earth's electric generator. Further computation can make clear this possibility.

If transferring valid for one occasion only, that can be made as the straight plasma cable 4 (Figure 5-4). For multi-applications the elliptic closed-loop plasma cable 6 is better. For permanent transmission the Earth must have a minimum two space towers (Figure 5-4). Many solar panels can be located on Moon and Moon can transfer energy to Earth.

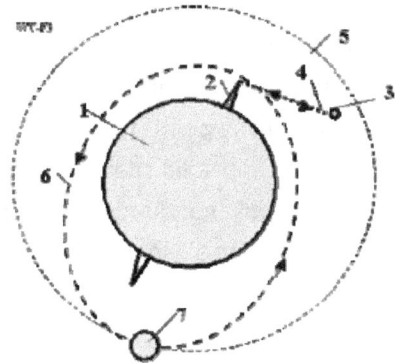

Figure 5-4. Transferring electric energy from Earth to satellite, Earth's International Space Station or to Moon (or back) by plasma cable. Notations: 1 - Earth, 2 - Earth's tower 100 km or more, 3 - satellite or Moon, 4 - plasma cable, 5 - Moon orbit, 6 - plasma cable to Moon, 7 - Moon.

3. Transferring Energy to Mars

The offered method may be applied for transferring energy to Mars including the case when Mars may be located in opposed place of Sun [2, Ch.6].

4. Plasma AB Magnetic Sail

Very interesting idea to build a gigantic plasma circle and use it as a Magnetic Sail (Figure 5-5) harnessing the Solar Wind. The computations show (see section "Macroproject") that the electric resistance of plasma cable is small and the big magnetic energy of plasma circle is enough for existence of a working circle in some years without external support. The connection of spaceship to plasma is also very easy. The space ship create own magnetic field and attracts to MagSail circle (if spacecraft is located behind the ring) or repels from MagSail circle (if spaceship located ahead of the ring). The control (turning of plasma circle) is also relatively easy. By moving the spaceship along the circle plate, we then create the asymmetric force and turning the circle. This easy method of building the any size plasma circle was discussed above.

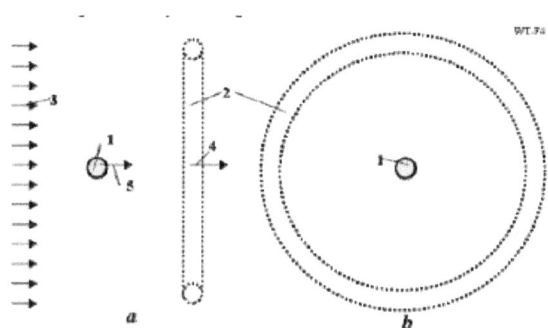

Figure 5-5. Plasma AB-MarSail. Notations: 1 - spaceship, 2 - plasma ring (circle), 3 - Solar wind, 4 - MagSail thrust, 5 - magnetic force of spaceship.

5. Wireless Transferring of Electric Energy in Earth

It is interesting the idea of energy transfer from one Earth continent to another continent without wires. As it is known the resistance of infinity (very large) conducting medium does not depend from distance. That is widely using in communication. The sender and receiver are connected by only one wire, the other wire is Earth. The author offers to use the Earth's ionosphere as the second plasma cable. It is known the Earth has the first ionosphere layer E at altitude about 100 km (Figure 5-6). The concentration of electrons in this layer reaches 5×10^4 1/cm3 in daytime and 3.1×10^3 1/cm3 at night. This layer can be used as a conducting medium for transfer electric energy and communication in any point of the Earth. We need minimum two space 100 km. towers. The cheap optimal inflatable, kinetic, and solid space towers are offered and researched by author in [1]-[4]. Additional innovations are a large inflatable conducting balloon at the end of the tower and big conducting plates in a sea (ocean) that would dramatically decrease the contact resistance of the electric system and conducting medium.

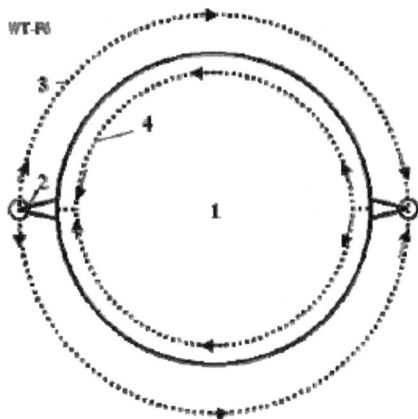

Figure 5-6. Using the ionosphere as conducting medium for transferring a huge electric energy between continents and as a large storage of the electric energy. Notations: 1 - Earth, 2 - space tower about 100 km of height, 3 - conducting E layer of Earth's ionosphere, 4 - back connection through Earth.

Conclusion

This new revolutionary idea - wireless transferring of electric energy in the hard vacuum of outer space is offered and researched. A rare plasma power cord as electric cable (wire) is used for it. It is shown that a certain minimal electric currency creates a compressed force that supports the plasma cable in the compacted form. Large amounts of energy can be transferred hundreds of millions of kilometers by this method. The requisite mass of plasma cable is merely hundreds of grams. It is computed that the macroprojects: transferring of hundreds of kilowatts of energy to Earth's International Space Station, transfer energy to Moon or back, transferring energy to a spaceship at distance of hundreds of millions of kilometers, transfer energy to Mars when it is on the other side of the Sun wirelessly. The transfer of colossal energy from one continent to another continent (for example, Europe to USA and back), using the Earth's ionosphere as a gigantic storage of electric energy, using the plasma ring as huge MagSail for moving of spaceships. It is also shown that electric currency in plasma cord can accelerate or slow various kinds of outer space apparatus.

6. Simplest AB-Thermonuclear Space Propulsion and Electric Generator*

The author applies, develops and researches mini-sized Micro- AB Thermonuclear Reactors for space propulsion and space power systems. These small engines directly convert the high speed charged particles produced in the thermonuclear reactor into vehicle thrust or vehicle electricity with maximum efficiency. The simplest AB-thermonuclear propulsion offered allows spaceships to reach speeds of 20,000 - 50,000 km/s (1/6 of light speed) for fuel ratio 0.1 and produces a huge amount of useful electric energy. Offered propulsion system permits flight to any planet of our Solar system in short time and to the nearest non-Sun stars by E-being or intellectual robots during a single human life period.

----
* See: http://arxiv.org/ftp/arxiv/papers/0706/0706.2182.pdf (2006) or [2, Ch.7].

Description of Innovations

The AB thermonuclear propulsion and electric generator are presented in Figure 6-1. As it is shown in [2] the minimized, or micro-thermonuclear reactor 1 generates high-speed charged particles 2 and neutrons that leave the reactor. The emitted charged particles may be reflected by electrostatic reflector, 4, or adsorbed by a semi-spherical screen 3; the neutrons may only be adsorbed by screen 3.

In *screen* of the AB-thermonuclear reactor (Figure 6-1a) the forward semi-spherical screen 3 adsorbed particles that move forward. The particles, 2, of the back semi-sphere move freely and produce the vehicle's thrust. The forwarded particles may to warm one side of the screen (the other side is heat protected) and emit photons that then create additional thrust for the apparatus. That is the *photon* AB-thermonuclear thruster.

In *reflector* AB-thermonuclear reactor (Figure 6-1b) the neutrons fly to space, the charged particles 5 are reflected the electrostatic reflector 4 to the side opposed an apparatus moving and create thrust.

The *screen-reflector* AB-thermonuclear reactor (Figure 6-1c) has the screen and reflector.

The *spherical* AB-propulsion-generator (Figure 6-1d) has two nets which stop the charged particles and produced electricity same as in [1, Ch. 17]. Any part 8 of the sphere may be cut-off from voltage and particles 9 can leave the sphere through this section and, thusly, create the thrust. We can change direction of thrust without turning the whole apparatus.

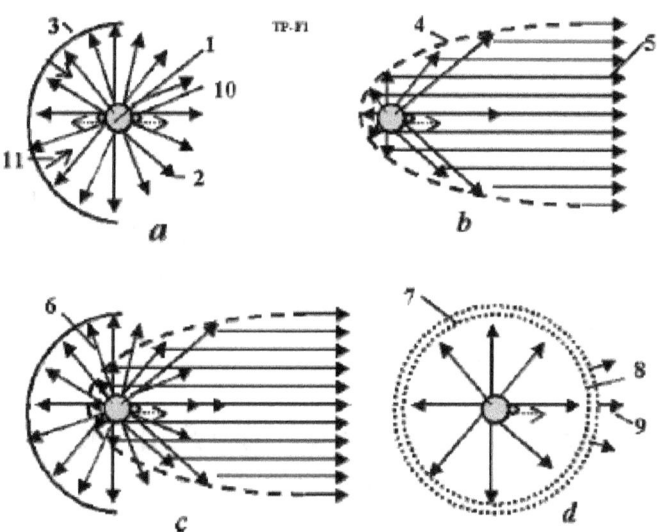

Figure 6-1. Types of the suggested propulsion and power system. (a) screen AB-thermonuclear propulsion and photon AB-thermonuclear propulsion ; (b) (electrostatic) reflector AB-thermonuclear propulsion; (c) screen-reflector AB-thermonuclear propulsion; (d) spherical AB-propulsion-generator. Notations: 1 - micro (mini) AB-thermonuclear reactor [15], 2 - particles (charged particles and neutrons) , 3 - screen for particles, 4 - electrostatic reflector; 5 - charged particles, 6 - neutrons, 7 - spherical net of electric generator, 8 - transparency (for charged particles) part of spherical net, 9 - charged particle are producing the thrust, 10 - electron discharger, 11 - photon radiation.

Conclusion

The author suggests the simplest maximally efficient thermonuclear AB-propulsion (and electric generators) based in the early offered size-minimized Micro-AB-thermonuclear reactor [2, Part B, Ch.1, p. 223]. These engines directly convert high-speed charged particles produced in thermonuclear reactor into vehicular thrust or onboard vehicle electricity resource. Offered propulsion system allows travel to any of our Solar System's planets in a short time as well as trips to the nearest stars by E-being or intellectual robot in during a single human life [2, Part C].

7. Electrostatic Linear Engine and Cable Space AB Launcher*

This is suggested a revolutionary new electrostatic engine. This engine can be used as a linear engine (accelerator), a strong space launcher, a high speed delivery system for space elevator, Earth-Moon, Earth-Mars, electrostatic train, levitation, conventional high voltage rotating engine, electrostatic electric generator, weapon, and so on. Author developed theory of this engine application and shows powerful possibility in space, transport and military industry. The projects are computed and show the good potential of the offered new concepts.

----

* This work is presented as Bolonkin's paper AIAA-2006-5229 for 42 Joint Propulsion Conference, Sacramento, USA, 9-12 July, 2006. Work is published in Aircraft Engineering and Aerospace Technology: An International Journal, Vol 78, #6, 2006, pp.502-508.
See also: http://arxiv.org/ftp/arxiv/papers/0705/0705.1943.pdf or [2, Ch.10].

Description of Electrostatic Linear Engine

The linear electrostatic engine (space accelerator) for launching of space ship [2] p.173 includes the following main parts (Figure 1): stator, thrust cable, charger of cable, high voltage electric alternating current line. As additional devices the engine can have a gas compaction, and vacuum pump.

The cable has a strong core (it keeps tensile stress - thrust) and dielectric cover contained electric charges. The conducting layer is very thin and we neglect its weight. A detailed linear engine (accelerator) for Cable Space Launcher is presented in Figure 7-2.

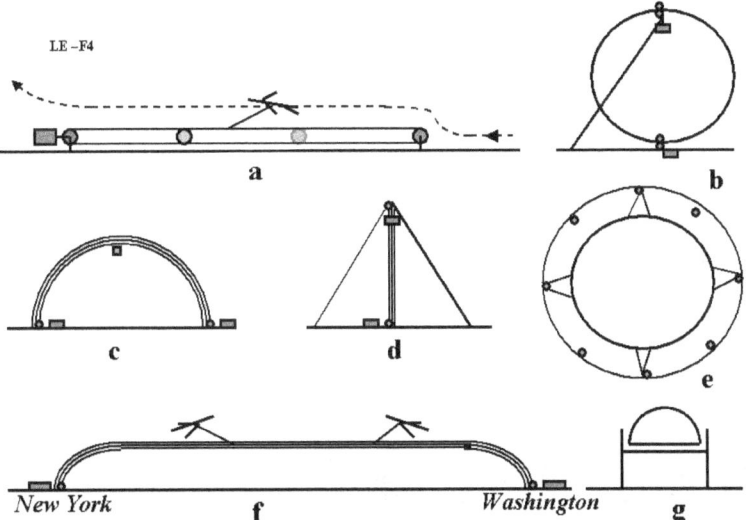

Figure 7-1. Installations needing the linear electrostatic engine. (a) Space cable launcher [2]; (b) Circle launcher; (c) Space keeper [1\; (d) Kinetic space tower[1]; (e) Earth round cable space keeper [1]; (f) Cable aviation [1]; (g) Levitation train [1].

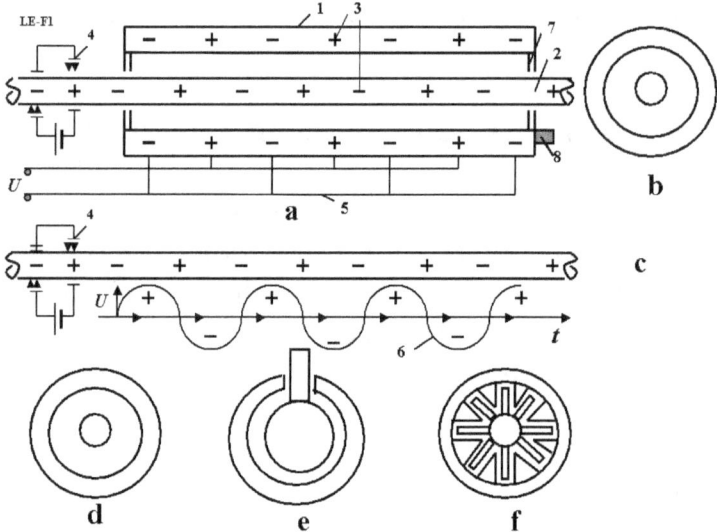

Figure 7-2. Electrostatic engine (accelerator) for Space Cable Launcher [23 -25]. (a) Engine (side view); (b) Engine (Forward view); (c) Running wave of voltage (charges) moves the charged cable; (d) - (f) Different cross-section areas of engine: (d) - conventional; (e) - for moving aircraft or space ship; (f) - for big thrust. *Notations*: 1 - stator of engine; 2 - thrust cable (rotor of engine); 3 - charges; 4 - recharger; 5 - high voltage line; 6 - alternating current (voltage); 7 - gas compaction; 8 - vacuum pump.

The engine works in the following mode. The cable has a set of stationary positive and negative charges. These charges can be restored if they are relaxed. Outer generator creates a running wave of voltage (charges) along stationary stator. This wave (charges) attracts the opposed charges in rotor (cable) and moves (thrusts) it.

Bottom and top parts of cable (or stator) have small different charge values. This difference creates a vertical electric field which supports the cable in suspended position inside stator non-contact bearing and zero friction. The cable position inside stator is controlled by electronic devices.

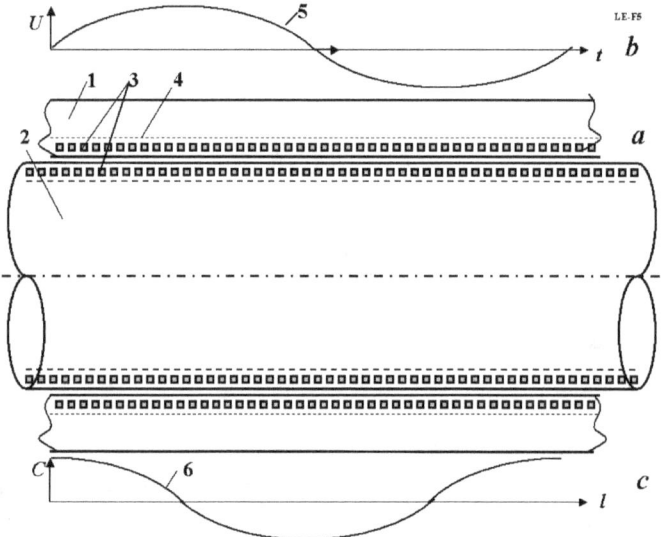

Figure 7-3. Detail electrostatic engine (accelerator) for Space Cable Launcher. (a) Accelerator, (b) Running voltage wave for stator, (c) Stationary charges into cable. *Notation*: 1 - stator; 2 - mobile rotor (cable); 3 -

charges; 4 - dielectric (isolator); 5 - running wave of voltage; 6 - curve of stationary cable charges. *U* is voltage, *C* is charge. Lagging between voltage wave of stator (b) and charges of mobile cable (c) is 90 degree.

Charges have toroidal form (row of rings) and located inside a good dielectric having high disruptive voltage. The stator toroids have conducting layer which allows changing the charges with high frequency and produces a running high voltage wave. The offered engine creates a large thrust (see computation below), reaches a very high (not limited) variable speed of cable (km/s), to change the moving of cable in opposed direction, to fix a cable in given position. The engine can also to work as high voltage electric generator when cable is braking or moved by mechanical force. The space elevator climber (and many other mobile apparatus) has constant charge, the cable (stator) has running charge. The weight of electric wires is small because the voltage is very high.

Conclusion

The offered electrostatic engine could find wide application in many fields of technology. That can decrease the launch cost from hundreds to thousands times. The electrostatic engine needs a very high voltage but this voltage is located in small area inside of installations and not dangerous to people. The current technology does not have another way for reaching a high speed except may be rockets. But rockets and rocket launches are very expensive and we do not know ways to decrease the cost of rocket launch thousand times.

8. AB Levitrons and their Applications to Earth's Motionless Satellites*

Author offers the new and distinctly revolutionary method of levitation in artificial magnetic field. It is shown that a very big space station and small satellites may be suspended over the Earth's surface and used as motionless radio-TV translators, telecommunication boosters, absolute geographic position locators, personal and mass entertainment and as planet-observation platforms. Presented here is the theory of big AB artificial magnetic field and levitation in it is generally developed. Computation of three macro-projects: space station at altitude 100 km, TV-communication antenna at height 500 m, and multi-path magnetic highway.

----

* Presented as Bolonkin's paper to http://arxiv.org Audust, 2007 (search "Bolonkin") or http://arxiv.org/ftp/arxiv/papers/0708/0708.2489.pdf or [2, Ch.12].

Innovations

The AB-Levitron uses two large conductivity rings with very high electric currency (fig.8-1). They create intense magnetic fields. Directions of electric currency are opposed one to the other and rings are repelling one from another. For obtaining enough force over a long distance, the electric currency must be very strong. The current superconductive technology allows us to get very high-density electric currency and enough artificial magnetic field in far space.

The superconductivity ring does not spend an electric energy and can work for a long time period, but it requires an integral cooling system because the current superconductivity materials have the critical temperature about 150-180 C.

For mobile vehicles the AB-Levitron can have a run-wave of magnetic intensity which can move the vehicle (produce electric currency), making it significantly mobile in the traveling medium.

The ring located in space does not need any conventional cooling—that defense from Sun and Earth radiations is provided by high-reflectivity screens (fig.3). However, that must have parts open to outer space for radiating of its heat and support the maintaining of low ambient temperature. For variable direction of radiation, the mechanical screen defense system may be complex. However, there are thin layers of liquid crystals that permit the automatic control of their energy reflectivity and transparency and the useful application of such liquid crystals making it easier for appropriate space cooling system. This effect is used by new man-made glasses which grow dark in bright solar light.

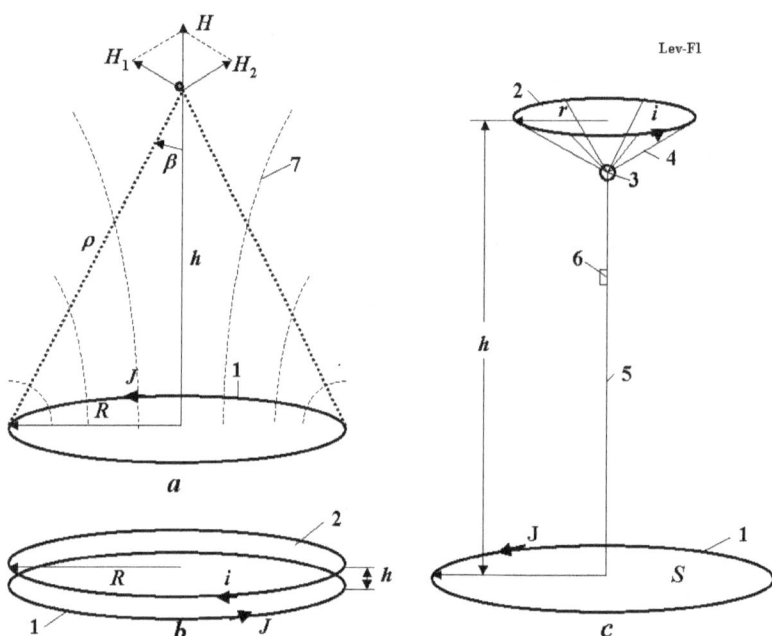

Figure 8-1. Explanation of AB-Levitron. (a) Artificial magnetic field; (b) AB-Levitron from two same closed superconductivity rings; (c) AB-Levitron - motionless satellite, space station or communication mast. Notation: 1- ground superconductivity ring; 2 - levitating ring; 3 - suspended stationary satellite (space station, communication equipment, etc.); 4 - suspension cable; 5 - elevator (climber) and electric cable; 6 - elevator cabin; 7 - magnetic lines of ground ring; R - radius of lover (ground) superconductivity ring; r - radius of top ring; h - altitude of top ring; H - magnetic intensity; S - ring area.

However, the present computation methods of heat defense are well developed (for example, by liquid nitrogen) and the induced expenses for cooling are small (fig. 8-2).

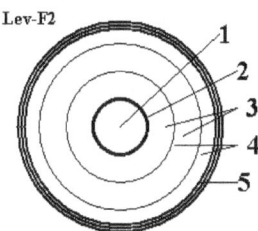

Figure 8-2. Cross-section of superconductivity ring. Notations: 1 - strong tube (internal part used for cooling of ring, external part is used for superconductive layer); 2 - superconductivity layer; 3 - vacuum; 4 – heat impact reduction high-reflectivity screens (roll of thin bright aluminum foil); 5 - protection and heat insulation.

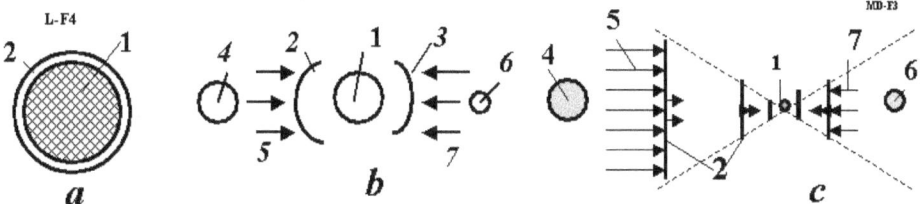

Figure 8-3. Methods of cooling (protection from Sun radiation) the superconductivity levitron ring in outer space. (*a*) Protection the ring by the super-reflectivity mirror [5]. (*b*) Protection by high-reflectivity screen (mirror) from impinging solar and planetary radiations. (*c*) Protection by usual multi-screens. Notations: 1 - superconductive wires (ring); 2 - heat protector (super-reflectivity mirror in Fig.3a and a usual mirror in Fig. 3c); 2, 3 – high-reflectivity mirrors (Fig. 3b); 4 - Sun; 5 -Sun radiation, 6 - Earth (planet); 7 - Earth's radiation.

The most important problem of AB-levitron is stability of top ring. The top ring is in equilibrium, but it is out of balance when it is not parallel to the ground ring. Author offers to suspend a load (satellite, space station, equipment, etc) lower then ring plate. In this case, a center of gravity is lower a summary lift force and system become stable.

Conclusion

We must research and develop these ideas. They may accelerate the technical progress and improve our life-styles. There are no known scientific obstacles in the development and design of the AB-Levitrons, levitation vehicles, high-speed aircraft, spaceship launches, low-aititude stationary tele-communication satellites, cheap space trip to Moon and Mars and other interesting destination-places in outer space.

9. Electrostatic Climber for Space Elevator and Launcher*

Here, the main author details laboratory and library research on the new, and intrinsically prospective, Electrostatic Space Elevator Climber. Based on a new electrostatic linear engine

---

* This work is presented as paper AIAA-2007-5838 for 43 Joint Propulsion Conference, Cincinnati, Ohio, USA, 9 -11 July, 2007. See also: http://arxiv.org/ftp/arxiv/papers/0705/0705.1943.pdf or [3, Ch.4].

previously offered at the 42nd Joint Propulsion Conference (AIAA-2006-5229) and published in "AEAT", Vol.78, No.6, 2006, pp. 502-508, the electrostatic climber described below can have any speed (and braking), the energy for climber movement is delivered by a light-weight high-voltage line into a Space Elevator-holding cable from Earth-based electricity generator. This electric line can be used for delivery electric energy to a Geosynchronous Space Station. At present, the best solution of the climber problem, announced by NASA, is very problematic.

Shown also, the linear electrostatic engine may be used nowadays as a realistic power space launcher. Two macro-projects illustrate the efficacy of these new devices.

Description of Electrostatic Linear Climber and Launcher

The linear electrostatic engine [2, Ch.10], climber, for Space Elevator includes the following main parts (Figure 9-1): plate (type) stator 1 (special cable of Space elevator), cylinders 3 inside having conducting layer (or net) (cylinder may be vacuum or inflatable film), conducting layer insulator, chargers (switches 6) of cable cylinders, high-voltage electric current line 6, linear rotor 7.

Linear rotor has permanent charged cylinder 4. As additional devices, the engine can have a gas-pressurizing capability and a vacuum pump [1-2].

The cable (stator) has a strong cover 2 (it keeps tensile stress - thrust/braking) and variable cylindrical charges contained dielectric cover (insulator). The conducting layer is very thin and we neglect its weight. Cylinders of film are also very light-weight. The charges can be connected to high-voltage electric lines 6 that are linked to a high-voltage device (electric generator) located on the ground.

The electrostatic engine works in the following mode. The rotor has a stationary positive charge. The cable has the variable positive and negative charges. These charges can be received by connection to the positive or negative high-voltage electric line located in cable (in stator). When positive rotor charge is located over given stator cylinder this cylinder connected by switch to positive electric lines and cylinder is charged positive charge but simultaneously the next stator cylinder is charged by negative charge. As result the permanent positive rotor charge repels from given positive stator charge and attracts to the next negative stator charge. This force moves linear rotor (driver). When positive rotor charge reaches a position over the negative stator charge, that charge re-charges to positive charge and next cylinder is connected to the negative electric line and then the whole cycle is repeated. To increase its efficiency, the positive and negative stator charges, before the next cycle, can run down through a special device, and their energy is returned to the electric line. It is noteworthy that the linear electrostatic engine can have very high efficiency!

Earth-constant potential generator creates a running single wave of charges along the stationary stator. This wave (charges) attracts (repel) the opposed (same) charges in rotor (linear driver) and moves (thrust or brake use) climber.

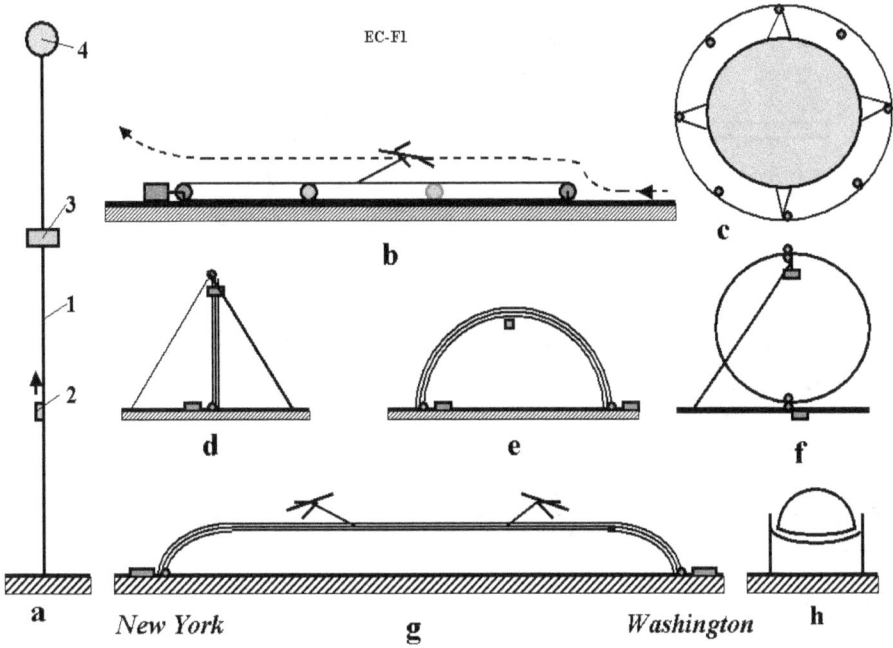

Figure 9-1. Installations needing the linear electrostatic engine. (a) Space elevator [36]. Notation: 1 - space elevator, 2 - climber, 3 - geosynchronous space station, 4 - balancer of space elevator. (b) Space cable launcher [3]; (b) Circle launcher [3]; (c) Earth round cable space keeper [38]; (d) Kinetic space tower [3]; (e - f) Space keeper [3]; (g) (f) Cable aviation [3]; (g) Levitation train [2-3].

The space launcher works same (Figure 9-2d, 2e). That has a stationary stator and mobile rotor (driver). The stationary stator (monorail) located upon the Earth's surface below the atmosphere. Driver is connected to space-aircraft and accelerates the aircraft to a needed speed (8 km/s and more) [2]. For increasing a thrust, the driver of the space launcher can have some charges (Figure 9-2d) separated by enough neutral non-charged stator cylinders.

Bottom and topmost parts of cable (or stator) have small different charge values. This difference creates a vertical electric field which supports the driver in its suspended position about the stator, non-contact bearing and zero friction. The driver position about the stator is controlled by electronic devices.

Charges have cylindrical form (row of cylinders), and are located within a good dielectric having high disruptive voltage. The cylinders have conducting layer which allows changing the charges with high frequency and produces a running high-voltage wave of charges. The engine creates a large thrust (see computation below), reaches a very high (practically speaking, virtually unlimited) variable speed of driver (km/s), to change the moving of driver in opposed direction, to fix a driver in selected given position. The electrostatic engine can also operate as a high-voltage electric generator when the climber (a cabin-style spaceship) is braking or is moved by some controlled mechanical force. The Space Elevator climber (and many other mobile apparatuses) has a constant charge; the cable (stator) has a running charge. The weight of electric wires is small, almost insignificant, because the voltage is very high.

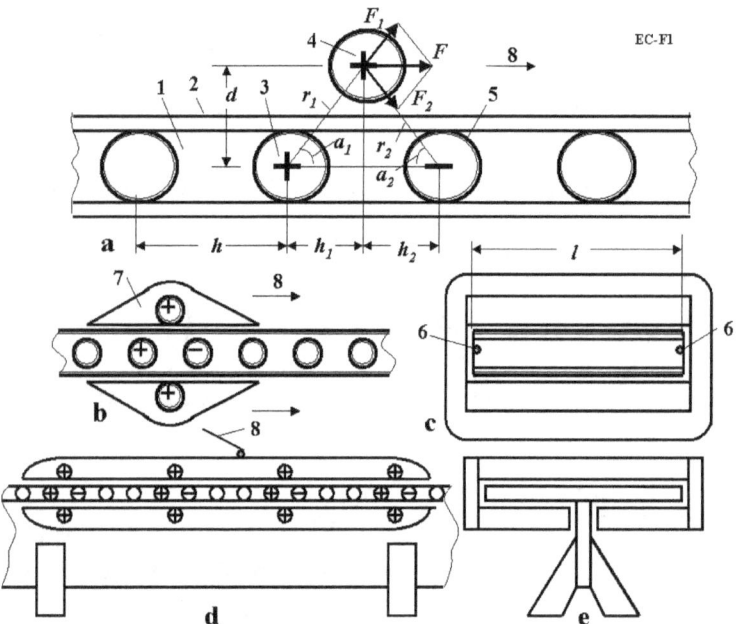

Figure 9-2. Electrostatic linear engine (accelerator) for Space Elevator and Space Launcher [23 -25, 40]. (a) Explanation of force in electrostatic engine; (b) Two cylindrical electrostatic engines for Space Elevator (here, a side view); (c) Two cylindrical electrostatic engine for Space Elevator (forward view); (d) Eight cylindrical electrostatic engine for Space Launcher (side view); (e) Eight cylindrical electrostatic engine for Space Launcher (forward view).
Notations: 1 - plate cable of Space Elevator with inserted variable cylindrical charges; 2 - part of cable-bearing tensile stress; 3 - insulated variable charging cylinder of stator; 4 - insulated permanent charging cylinder of rotor; 6 – high-voltage wires connected with Earth's generator and switch; 7 - mobile part of electrostatic linear engine; 8 - cable to space aircraft.

Conclusion

Presently, the suggested space climber is the single immediately buildable high-efficiency transport system for a space elevator. The electromagnetic beam transfer energy is very complex, expensive and has very low efficiency, especially at a long-distance from divergence of electromagnetic beam. The laser has similar operational disadvantages. The conventional electric line, equipped with conventional electric motor, is very heavy and decidedly unacceptable for outer space.

The offered electrostatic engine could find wide application in many fields of technology. It can drastically decrease the monetary costs of launch by hundreds to thousands of times. The electrostatic engine needs a very high-voltage but this voltage, however, is located in a small area inside of installations, and is not particularly dangerous to person living or working nearby. Currently used technology does not have any other way for reaching a high speed except by the use of rockets. But crewed or un-crewed rockets, and rocket launches for expensive-to-maintain-and-operate Earth bases, are very expensive and space businesses do not know ways to cut the cost of rocket launch by hundreds to thousands of times.

10. AB-Space Propulsion*

On 4 January 2007 the article "Wireless Transfer of Electricity in Outer Space" in http://arxiv.org was published wherein was offered and researched a new revolutionary method of

---
* Presented in http://arxiv.org of Cornel University (1 March, 2008). http://arxiv.org/ftp/arxiv/papers/0803/0803.0089.pdf , or [3] Ch.10.

transferring electric energy in space. In next article (see http://arxiv.org) was developed the theory of new engine.

That Chapter describes a new engine which produces a large thrust without throwing away large amounts of reaction mass (unlike the conventional rocket engine). A sample computed project shows the big possibilities opened by this new "AB-Space Engine". The AB-Space Engine gets the energy from ground-mounted power; a planet's electric station can transfer electricity up to 1000 millions (and more) of kilometers by plasma wires. Author shows that AB-Space Engine can produce thrust of 10 tons (and more). That can accelerate a spaceship to some thousands of kilometers/second. AB-Space Engine has a staggering specific impulse owing to the very small mass expended. The AB-Space Engine reacts not by expulsion of its own mass (unlike rocket engine) but against the mass of its planet of origin (located perhaps a thousand of millions of kilometers away) through the magnetic field of it's plasma cable. For creating this plasma cable the AB-Space Engine spends only some kg of hydrogen.

Offered Innovations and Brief Descriptions

1. **Transfer of electricity by plasma cable.** The author offers a series of innovations that may solve the many macro-problems of transportation, energy and thrust in space. Below are some of them.

   1. Transfer of electrical energy in outer space using a conductive cord from plasma. Author has solved the main problem - how to keep the plasma cord from dissipation, and in compressed form. He has developed the theory of space electric transference, made computations that show the possibility of realization for these ideas with existing technology. The electric energy may be transferred for hundreds millions of kilometers in space (including Moon and Mars) [1].
   2. Method of construction of space electric lines and electric devices.
   3. Method of utilization and tapping of the plasma cable electric energy.
   4. Two methods of converting the electric energy to impulse (thrust) motion of a spacecraft (these two means are utilization of the magnetic field and of the kinetic energy of ions and electrons of the electric current).
   5. Design of a triple electrostatic mirror (plasma reflector), which can reflect the plasma flow [1].

Below are some succinct descriptions of some constructions made possible by these revolutionary macro-engineering ideas.

1. Transferring electric energy in Space. The electric source (generator, station) is connected to the distant location in space by two artificially generated rarefied plasma cables (Figure 10-1a). These cables can be created by a plasma beam [1, 2] sent from the Moon, Earth mounted super high tower, or from a space station in low Earth orbit, or a local base at the target location. If the plasma beam is sent remotely from the Earth, a local reflector station is required at the target site or at a third location to turn the circuit back toward its' starting point and closure.

The plasma cable, in radial direction, may also be constructed of ultra-cold plasma.

The plasma cable is self-supported in cable form by the magnetic field created by the electric current going through the plasma cable. The axial electric current produces an contracting magnetic pressure opposed to an expansive gas dynamic plasma pressure (the well-known theta-pinch effect)(Figure 10-2b). The plasma has a good conductivity (equal to that of silver and more) and the plasma cable can have a very big cross-section area (up to thousands of square meters cross-section). The plasma conductivity does not depend on its' density. That way the plasma cable has no large resistance although the length of plasma cable is hundreds of millions of kilometers. The needed minimum electric current is derived from parameters of a plasma cable researched in the theoretical section of this article.

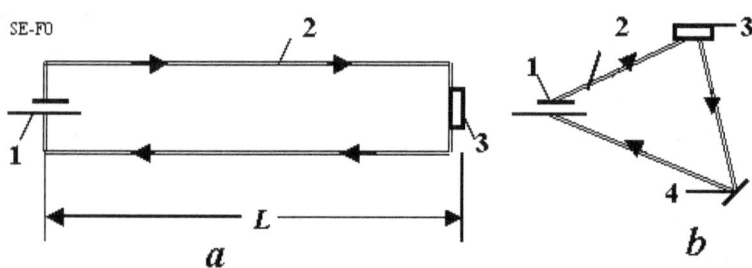

Figure 10-1. Long distance plasma transfer electric energy and thrust in outer space. *a* - plasma transfer with parallel plasma cable, *b* - plasma transfer with triangular (three-wire) plasma cable. *Notations*: 1 - current source (generator), 2 - plasma wire (cable), 3 - spaceship, orbital station or other energy destinations, 4 - plasma reflector located at planet, asteroid or space station.

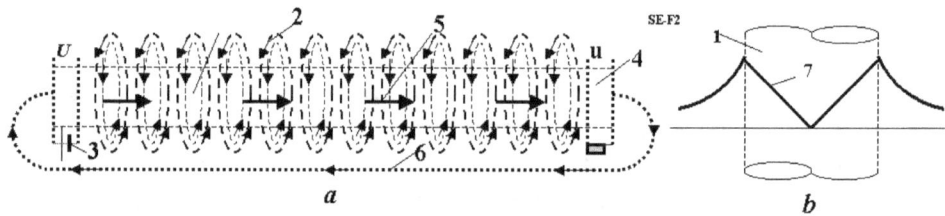

Figure 10-2. *a*. A plasma cable supported by its' own magnetic field, *b*. Magnetic intensity into and out of plasma cable. *Notations*: 1 -plasma cable, 2 - compressing magnetic field, 3 - electric source, 4 - electric receiver, 5 - electric current, 6 - back plasma line; 7 – magnetic intensity into and out of plasma cable.

The parallel cables having opposed currents repel each other (Figure 10-2a)(by magnetic force). This force may be balanced by attractive electric force if we charge the cables by electric charges (see theoretical section). They also can be separated by a special plasma reflector as it shown in figures 10-2b. The electric line can be created and exist independently. The spaceship connects to this line at a suitable point. By altering the diameter and direction of the plasma cable we can supply energy to a spacecraft. Though we must supply energy to accelerate the spacecraft we can also regenerate energy by its braking activity. At any time, the spaceship can disconnect from the line and can exist without line support (propulsion, electricity, etc). The apparatus can hook up to or disconnect from the plasma cable at will. But breaking (loss of continuity) of the plasma cable itself destroys the plasma cable line to the remote location! We must have additional (parallel) plasma lines and apparatus must disconnect from a damaged or occulted (for example on the far side of a remote planet) plasma line and connect to another line to keep the connection in existence. The same situation is true in a conventional electric net. The apparatus can also restore the damaged part of plasma line by own injected plasma, but the time for repairing is limited (by tens of minutes or a few hours). The original station can also to send the plasma beam which connects the ends of damaged part of the line.

The electric tension (voltage) in a plasma cable is between two ends (for example, as cathode- anode) of the conductor in the issuing electric station (electric generator) [1-4]. The plasma cable current has two flows: Electron (negative) flow and opposed ions (positive) flow in the same cable.

These flows create an electric current. (In metal we have only electron flow, in liquid electrolytes we have ions flowing).

The author offers methods (for extraction and inserting) of energy from the plasma electric cable (Figure 3) by customer (spacecraft, other energy destination or end user).

The double net can accelerate the charged particles and insert energy into plasma cable (fig, 3a) or brakes charged particles and extract energy from electric current (figure 3b). In the first case the two nets create the straight electrostatic field, in the second case the two nets create the opposed electrostatic field in plasma cable (resistance in the electric cable [1- 4]) (figures 10-2, 10-3c). This apparatus resistance utilizes the electric energy for the spaceship or space station. In the second case the charged particles may be collected into a set of thin films and emit (after utilization in apparatus) back into continued plasma cable (see [1- 4]).

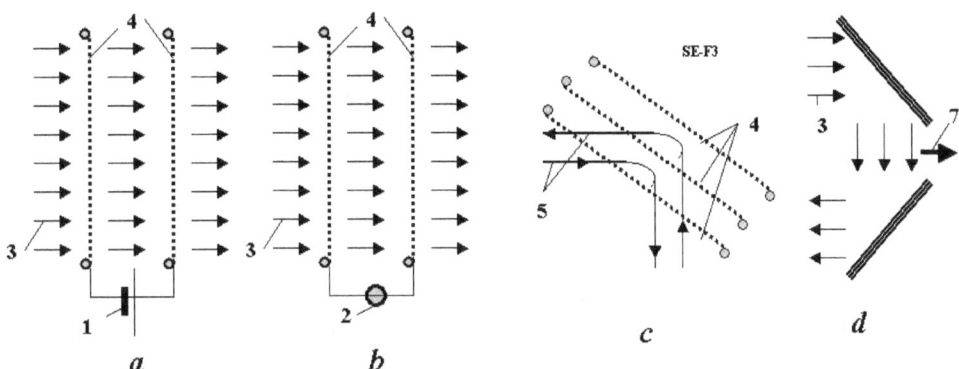

Figure 10-3. Getting and inserting in (off) plasma cable the energy and turning of plasma cable. *a* – inserting electric energy into plasma cable by means of two thin conducting nets or films; *b* - getting the energy from plasma cable by means of two thin conducting nets or films; *c* – offered triple net plasma reflector; *d* – double triple net plasma reflectors - the simplest AB thruster. *Notations*: 1 - spaceship or space station, 2 – receiver of energy, 3 - plasma cable, 4 - electrostatic nets, 5 – two opposed flows of charged plasma particles (negative and positive: electron and ions), 7 – thrust of AB-Space Engine.

Figure 10-3c presents the plasma beam reflector [1-4]. That has three charged nets. The first and second nets reflect (for example) positive particles, the second and third nets reflected the particles having an opposed charge.

Figure 10-4 shows the different design the plasma cable in space.

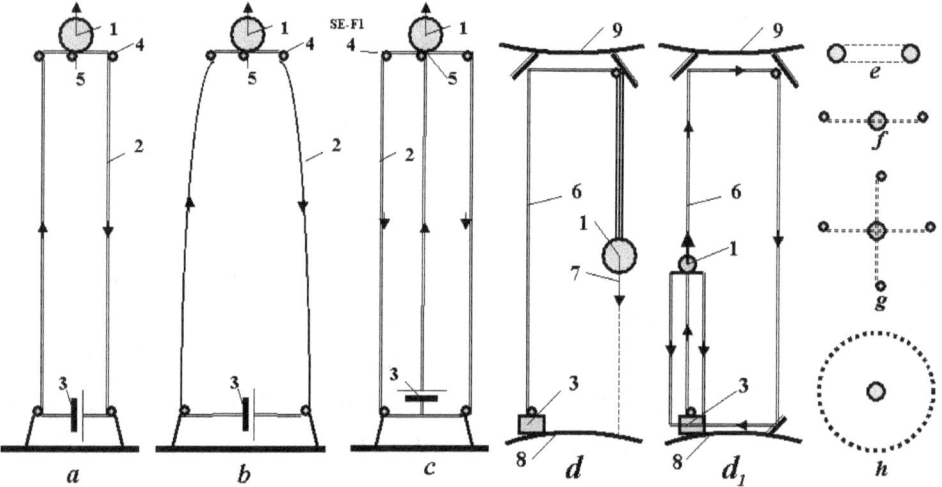

Figure 10-4. Transfer electricity and thrust by AB-Space Engine: *a*. Two plasma parallel cables; *b*. Curved cable; *c.* Plasma multi-cables; *d*. Transfer of back thrust through planet or asteroid; d_1. Using of ready plasma line; *e – h*. Forms of straight and back plasma cables (cross-sections of cables). *Notations*: 1 – Space ship; 2 – plasma cable; 3 – source electricity; 4 - plasma injector; 5 – user of energy; 6 – double plasma line; 7 - thrust; 8 – Earth; 9 – planet or asteroid.

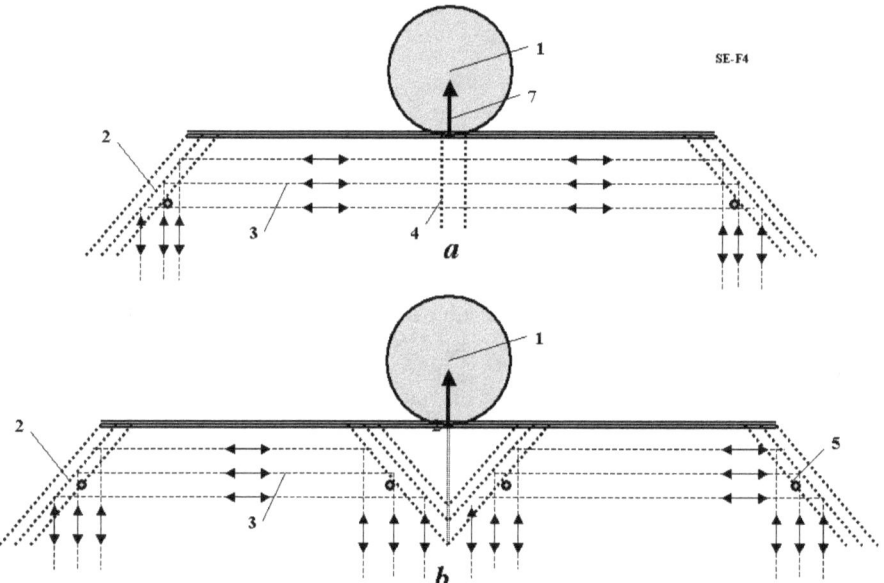

Figure 10-5. Some versions of AB-Space Engine (thruster). a. two cable AB-Space Engine; b. Three cable AB-Space Engine. Notations; 1 – space ship; 2 – offered special (three nets) electrostatic reflector; 3 – plasma cable; 4 – receiver or source of energy; 5 – injector of plasma, 7 – thrust.

Figure 10-4a shows two plasma parallel cables. Figure 4b two shows plasma parallel cables of a curved form of line. Figure 10-4c presents three plasma parallel cables, one to space ship and two for back (return) current. Figure 10-4d shows the transfer of the reverse impulse (or braking) thrust to space ship through planet or asteroid. Figures 10-5e-h shows the different forms of the straight and back plasma cables (cross-section of cables).

2. AB-Space Engine. The offered simplest AB-Space Engine is shown in Figure 10-4d and more details in Figures 10-5, 6. That includes two new triple electrostatic reflectors 2 which turn the plasma cables' (flow), (electric current 3) in back direction. The engine may also contain (optional) the plasma injectors 5 and electric generator (user) 4.

As feed material for the plasma may be used hydrogen gas, as plasma reflector may be used three conductivity nets connected to voltage sources, as generator - the double conductivity nets located into plasma flow and connected to voltage sources or users.

The other design of AB-Space Engine is shown in figure 10-6b. Here the central plasma flow divides in two side flows which go back to the electric station.

The AB-Space Engine works as follows. The electric current (voltage) produced by electric station (that may be located far from AB-Space Engine, for example, in orbit around the Earth or mounted on the Moon, Phobos or another space body) transfers by plasma cable to the AB-Space Engine. The power of the electric current in the plasma produces the power plasma flow of electrons and ions. The engine turns back the plasma flow (electric current) and returns it to the source electric station by the other plasma cable. The magnetic and centrifugal forces appear at the point of turning from outgoing to ingoing plasma paths place and create the thrust which can be used for movement (acceleration, braking) the space apparatus (or conventional vehicle or projectile).

Long-time readers of proposed space drive papers may suspect something fishy here. Don't worry: The AB-Space Engine doesn't violate Newton's third law of action and reaction. The AB-Space Engine reacts against the (planet or station mounted) electric station which may be located hundreds of millions of kilometers away! No other engine has the same capability.

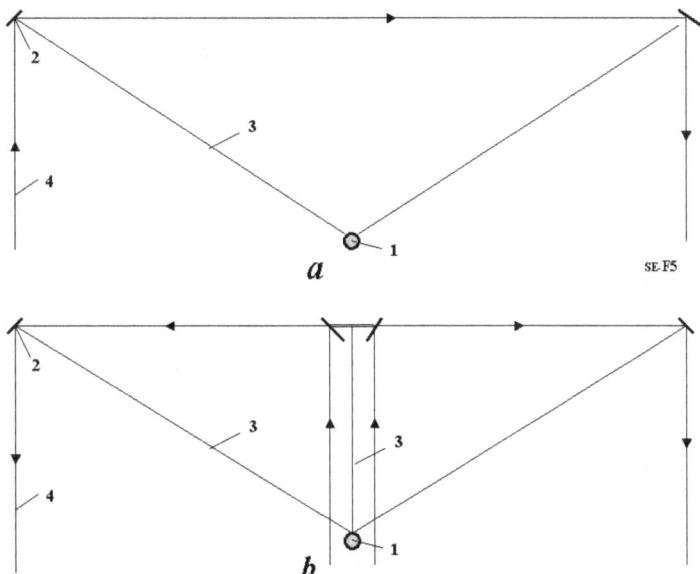

Figure 10-6. Some versions of widely (many km) AB-Space Engine (thruster). a. two cable AB-Space Engine; b. Three cable AB-Space Engine. Notations; 1 – space ship; 2 – offered special (three nets) electrostatic reflector; 3 – thin space cable connected the ship and reflector; 4 – plasma cable.

Your attention is also directed to the following differences between a railgun and an AB-Space Engine:

1) The railgun uses SOLID physical rails for delivery the electric current to conductivity projectile. These are easily damaged by huge electric current. The AB-Space Engine uses flexible plasma cables which can self-repair.
2) The railgun uses the rails which are of fixed construction (unalterable) and a spacecraft so launched can move solely in the rail direction. The AB-Space Engine creates the plasma cable in the course of apparatus movement and can select and change the apparatus' future direction.
3) Even a theoretical railgun girdling the globe of the Moon in vacuum (for star probe launch) would have a possible length of only some kilometers (as any solid construction). The plasma electric line (used byAB-Space Engine) can have a length (an acceleration path) of millions of kilometers (and thus may someday power manned craft on missions to near interplanetary space).

Discussion

Advantages of AB-engine

1) The offered AB-Space Engine is very light, simple, safe, and reliable with comparison to any likely nuclear engine.
2) The AB-Space Engine has a gigantic 'virtual specific impulse', being more capable of realistic operation in a projectable near-future environment, than virtually any proposed means of thermonuclear or light-propulsion scheme the author is aware of.
3) The AB-Space Engine can accelerate a near-term space apparatus to very high speed (approaching light speed). At present time this is the single real method to be able to approach this 'ultimate' velocity.
4) At least part of the needed injected plasma cable mass and nearly all of the energy needed (and the cooling facilities needed to maintain that energy supply) can be from the planet-bound energy-supplying station, further improving the on-board ship 'mass ratio'.
5) The AB-Space Engine can use far cheaper energy from a planet-bound electric station.

The offered ideas and innovations may create a jump in space and energy industries. Author has made initial base researches that conclusively show the big possibilities offered by the methods and installations proposed. Further research and testing are necessary. Those tests are not expensive. As that is in science, the obstacles can slow, even stop, applications of these revolutionary innovations. For example, the plasma cable may be unstable. The instability mega-problem of a plasma cable was found in tokomak R&D, but it is successfully solved at the present time. The same method (rotation of plasma cable) can be applied in our case.

The other problem is production of the plasma cable in Earth's atmosphere. This problem may be sidestepped by operations from a suitably high super-stratospheric tower such as outlined in others of the author's works, or is no problem at all if the electric station of the plasma cables' origin is located on the Moon [3].

The author has ideas on how to solve this problem with today's technologies and to use the readily available electric stations found on this planet Earth. Inquiries from serious parties are invited.

Summary

This new revolutionary idea – The AB-Space Engine and wireless transferring of electric energy in the hard vacuum of outer space – is offered and researched. A rarefied plasma power cord in the function of electric cable (wire) is used for it. It is shown that a certain minimal electric current creates a compression force that supports and maintains the plasma cable in its compacted form. Large amounts of energy can be transferred hundreds of millions of kilometers by this method. The requisite mass of plasma cable is merely hundreds of grams (some kg). A sample macroproject is computed: An AB-Space Engine having thrust = 10 tons. It is also shown that electric current in plasma cord can accelerate or slow various kinds of outer space apparatus.

11. Wireless Transfer of Electricity from Continent to Continent*

Author offers collections from his previous research of the revolutionary new ideas: wireless transferring electric energy in long distance – from one continent to other continent through Earth ionosphere and storage the electric energy into ionosphere. Early he also offered the electronic tubes as the method of transportation of electricity into outer space and the electrostatic space 100 km towers for connection to Earth ionosphere.
Early it is offered connection to Earth ionosphere by 100 km solid or inflatable towers. There are difficult for current technology. In given work the research this connection by thin plastic tubes supported in atmosphere by electron gas and electrostatic force. Building this system is cheap and easy for current technology.

The computed project allows to estimate the possibility of the suggested method.

----
*See: http://www.scribd.com/doc/42721638/ or [4].

Wireless transferring of electric energy in Earth.

It is interesting the idea of energy transfer from one Earth continent to another continent without wires. As it is known the resistance of infinity (very large) conducting medium does not depend from distance. That is widely using in communication. The sender and receiver are connected by only one wire, the other wire is Earth. The author offers to use the Earth's ionosphere as the second plasma cable. It is known the Earth has the first ionosphere layer E at altitude about 100 km (Fig. 1). The concentration of electrons in this layer reaches 5×10^4 1/cm^3 in daytime and 3.1×10^3 1/cm^3 at night (Fig. 11-1). This layer can be used as a conducting medium for transfer electric energy and communication in any point of the Earth. We need minimum two space 100 km. towers (Fig. 11-2). The cheap optimal inflatable, kinetic, and solid space towers are offered and researched by author in [4]. Additional innovations are a large inflatable conducting balloon at the end of the tower and big conducting plates in a sea (ocean) that would dramatically decrease the contact resistance of the electric system and conducting medium.

Theory and computation of these ideas are presented in Macroprojects section [4].

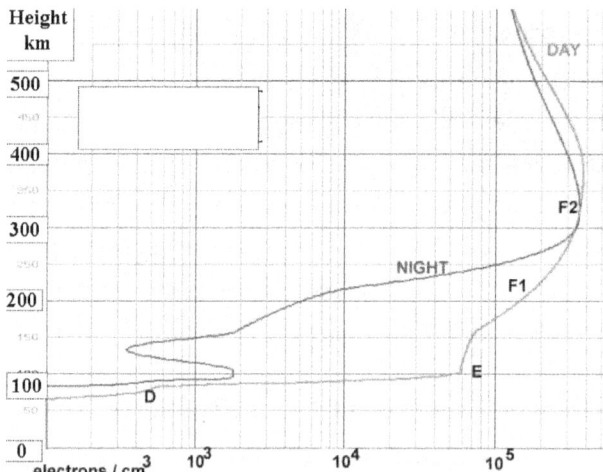

Fig. 11-1. Consentration/cm³ of electrons (= ions) in Earth's atmosphere in the day and night time in the D, E, F1, and F2 layers of ionosphere.

However the solid 100 km space towers are very expensive. Main innovation in this work is connection to ionosphere by cheap film tube filled by electron gas.

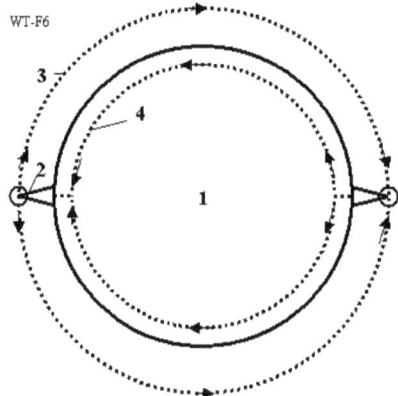

Fig. 11-2. Using the ionosphere as conducting medium for transferring a huge electric energy between continents and as a large storage of the electric energy. Notations: 1 - Earth, 2 - space tower (or electron tube) about 100 km of height, 3 - conducting *E* layer of Earth's ionosphere, 4 - back connection through Earth.

Electronic tubes

The author's first innovations in electrostatic applications were developed in 1982-1983 [1]-[4], [3, p.497].Later the series articles of this topic were published in [2, Ch.6]. In particular, in the work [4-5] was developed theory of electronic gas and its application to building (without space flight!) inflatable electrostatic space tower up to the stationary orbit of Earth's satellite (GEO) [2, Ch.11].

In given work this theory applied to special inflatable electronic tubes made from thin insulator film. It is shown the charged tube filled by electron gas is electrically neutral, that can has a high internal pressure of the electron gas.

The main property of AB electronic tube is a very low electric resistance because electrons have small friction on tube wall. (In conventional solid (metal) conductors, the electrons strike against the immobile ions located in the full volume of the conductor.). The abnormally low electric resistance was found along the lateral axis only in nanotubes (they have a tube structure!). In theory, metallic nanotubes can have an electric current density (along the axis) more than 1,000 times greater than metals such as silver and copper. Nanotubes have excellent heat conductivity along axis up 6000 W/m·K. Copper, by contrast, has only 385 W/m·K. The electronic tubes explain why there is this effect. Nanotubes have the tube structure and electrons can free move along axis (they have only a friction on a tube wall).

More over, the moving electrons produce the magnetic field. The author shows - this magnetic field presses against the electron gas. When this magnetic pressure equals the electrostatic pressure, the electron gas may not remain in contact with the tube walls and their friction losses. The electron tube effectively becomes a superconductor for any surrounding temperature, even higher than room temperature! Author derives conditions for it and shows how we can significantly decrease the electric resistance.

Description, Innovations, and Applications of Electronic tubes.

An electronic AB-Tube is a tube filled by electron gas (fig.11-3). Electron gas is the lightest gas known in nature, far lighter than hydrogen. Therefore, tubes filled with this gas have the maximum possible lift force in atmosphere (equal essentially to the lift force of vacuum). The applications of electron gas are based on one little-known fact – the electrons located within a cylindrical tube having a positively charged cover (envelope) are in neutral-charge conditions – the total attractive force of the positive envelope plus negative contents equals zero. That means the electrons do not adhere to positive charged tube cover. They will freely fly into an AB-Tube. It is known, if the Earth (or other planet) would have, despite the massive pressures there, an empty space in Earth's very core, any matter in this (hypothetical!) cavity would be in a state of weightlessness (free fall). All around, attractions balance, leaving no vector 'down'.

Analogously, that means the AB-Tube is a conductor of electricity. Under electric tension (voltage) the electrons will collectively move without internal friction, with no vector 'down' to the walls, where friction might lie. In contrast to movement of electrons into metal (where moving electrons impact against a motionless ion grate). In the AB-Tube we have only electron friction about the tube wall. This friction is significantly less than the friction electrons would experience against ionic structures—and therefore so is the electrical resistance.

When the density of electron gas equals $n = 1.65 \times 10^{16}/r$ $1/m^3$ (where r is radius of tube, m), the electron gas has pressure equals atmospheric pressure 1 atm (see research below). In this case the tube cover may be a very thin—though well-sealed-- insulator film. The outer surface of this film is charged positively by static charges equal the electron charges and AB-Tube is thus an electrically neutral body.

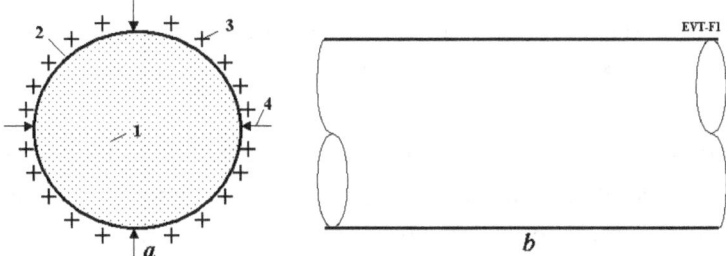

Fig. 11-3. Electronic vacuum AB-Tube. *a*) Cross-section of tube. *b*) Side view. *Notation*: 1 – Internal part of tube filled by free electrons; 2 – insulator envelope of tube; 3 – positive charges on the outer surface of envelope (over this may be an additional film-insulator); 4 – atmospheric pressure.

Moreover, when electrons move into the AB-Tube, the electric current produces a magnetic field (fig. 11-4). This magnetic field compresses the electron cord and decreases the contact (and friction, electric resistance) electrons to tube walls. In the theoretical section is received a simple relation between the electric current and linear tube charge when the magnetic pressure equals to electron gas pressure $i = c\tau$ (where i is electric current, A; $c = 3 \times 10^8$ m/s – is the light speed; τ is tube linear electric charge, C/m). In this case the electron friction equals zero and AB-Tube becomes **superconductive at any outer temperature**. Unfortunately, this condition requests the electron speed equals the light speed. It is, however, no problem to set the electron speed very close to light speed. That means we can make the electric conductivity of AB-Tubes very close to superconductivity almost regardless of the outer temperature.

Summary

This new revolutionary idea - wireless transferring of electric energy in long distance through the ionosphere or by the electronic tubes is offered and researched. A rare plasma power cord as electric cable (wire) is used for it. It is shown that a certain minimal electric currency creates a compressed force that supports the plasma cable in the compacted form. Large amounts of energy can be transferred many thousands of kilometers by this method. The requisite mass of plasma cable is merely hundreds of grams. It is computed that the macroproject: The transfer of colossal energy from one continent to another continent (for example, Europe to USA and back), using the Earth's ionosphere as a gigantic storage of electric energy.

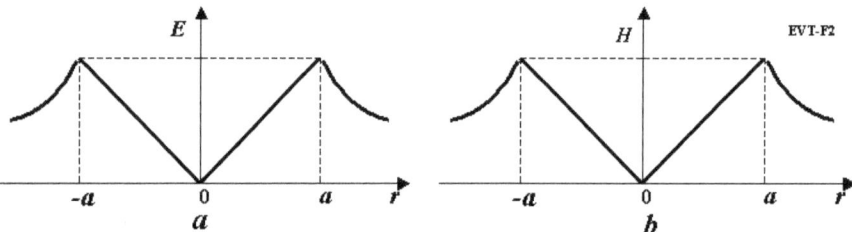

Fig. 11-4. Electrostatic and magnetic intensity into AB-Tube. *a*) Electrostatic intensity (pressure) via tube radius. *b*) Magnetic intensity (pressure) from electric current versus rube radius.

12. Magnetic Space Launcher*

A method and facilities for delivering payload and people into outer space are presented. This method uses, in general, engines located on a planetary surface. The installation consists of a space apparatus, power drive stations, which include a flywheel accumulator (for storage) of energy, a variable reducer, a powerful homopolar electric generator and electric rails. The drive stations accelerate the apparatus up to hypersonic speed.

The estimations and computations show the possibility of making this project a reality in a short period of time (for payloads which can tolerate high *g*-forces). The launch will be very cheap at a projected cost of $3 - $5 per pound. The authors developed a theory of this type of the launcher.

---

*Presented as paper AIAA-2009-5261 to 45th AIAA Joint Propulsion Conference, 2-5 August 2009, Denver, CO, USA. See also http://www.scribd.com/doc/24051286/ or [4].

Description of Suggested Launcher

Brief Description. The installation includes (see notations in Figs. 12-1, 12-2): a gun, two electric rails 2, a space apparatus 3, and a drive station 4 (fig. 12-1). The drive station includes: a homopolar electric generator 1 (fig. 12-2), a variable reducer 3, a fly-wheel energy storage 5, an engine 6, and master drive clutches 2, 4, 6.

The system works in the following way:

The engine 7 accelerates the flywheel 5 to maximum safe rotation speed. At launch time, the fly wheel connects through the variable reducer 3 to the homopolar electric generator 1 which produces a high-amperage current. The gas gun takes a shot and accelerates the space apparatus "c" (fig. 12-3) up to the speed of 1500 – 2000 m/s. The apparatus leaves the gun and gains further motion on the rails 2 (fig. 12-3) where its body turns on the heavy electric current from the electric generator. The magnetic force of the electric rails accelerates the space apparatus up to speeds of 8000 m/s. (or more). The initial acceleration with a gas gun can decrease the size and cost of the installation when the final speed is not high. The gas gun cannot produce a projectile speed of more than about 2000 m/s. The railgun does not have this limit, but produces some engineering problems such as the required short (pulsed) gigantic surge of electric power, sliding contacts for some millions of amperes current, storage of energy, etc.

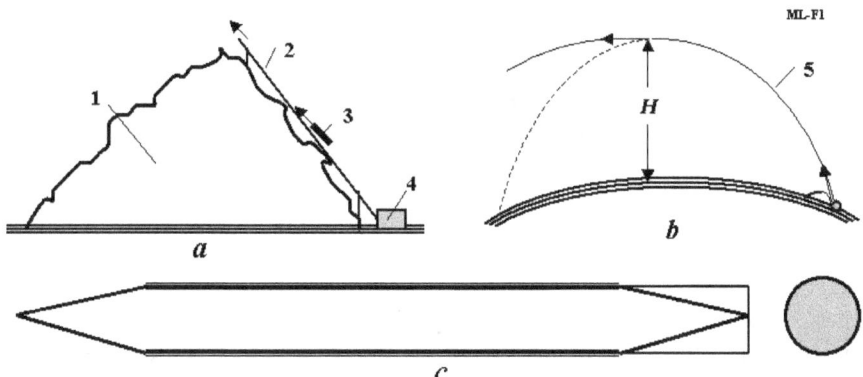

Fig. 12-1. Magnetic Launcher. (*a*) Side view; (*b*) Trajectory of space apparatus; (*c*) Hypersonic apparatus. *Notations*: 1 – hill (side view); 2 – railing; 3 – shell; 4 – drive station; 5 – space trajectory.

The current condensers have a small electric capacity 0.002 MJ/kg ([3, p.465)]. We would need about 10^{10} J energy and 5000 tons of these expensive condensers. The fly-wheels made of cheap artificial fiber have capacity about 0.5 MJ/kg ([3, p.464]). The need mass of fly-wheel is decreased to a relatively small 25 – 30 tons. The unit mass of a fly-wheel is significantly cheaper then unit mass of the electric condenser.

The offered design of the magnetic launcher has many innovations which help to overcome the obstacles afforded by a conventional railgun. Itemizing some of them:

1. Fly-wheels from artificial fiber.
2. Small variable reducer with smooth change of turns and high variable rate.

3. Multi-stage monopolar electric generator having capacity of producing millions of amperes and a variable high voltage during a short time.
4. Sliding mercury (gallium) contact having high pass capacity.
5. Double switch having high capacity and short time switching.
6. Special design of projectile (conductor ring) having permanent contact with electric rail.
7. Thin (lead) film on projectile contacts that improve contact of projectile body and the conductor rail.
8. Homopolar generator has magnets inserted into a disk (wheel) form. That significantly simplifies the electric generator.
9. The rails and electric generator can have internal water-cooling.
10. The generator can return rotation energy back to a flywheel after shooting, while rails can return the electromagnetic energy to installation. That way a part of shot energy may be returned. This increases the coefficient of efficiency of the launch installation.

The fly-wheel has a disadvantage in that it decreases its' turning speed when one spends its energy. The prospective space apparatus and space launcher needs, on the contrary, an increase of voltage for accelerating the payload. The homopolar generator really would like to increase the number of revolutions thus increasing the voltage. The offered variable reducer approaches this ideal, keeping constant or even increasing the speed of rotation of the electric generator. In addition, the multi-stage electric generator can additionally increase its' voltage by chaining (concatenation of turning on in series mode) its stages or sections.

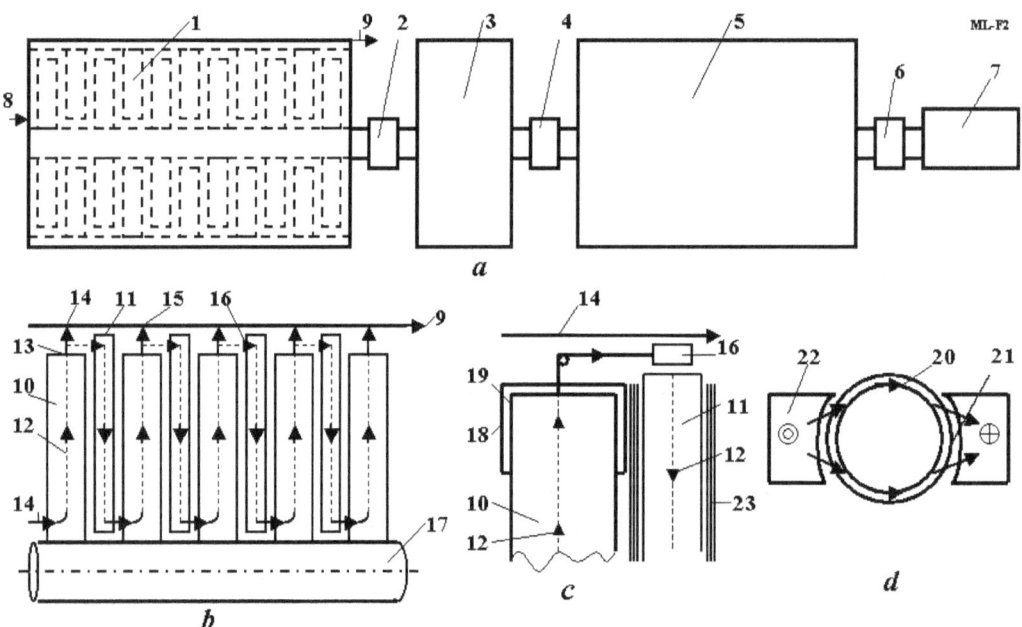

Fig. 12-2. Drive station. (*a*) Main components of drive station; (*b*) Rotors and connection disks (wheels); (*c*) Association of rotor and connection disk; (*d*) Association of shell and electric rails (plough or sled). *Notations:* 1 – Electric homogenerator; 2, 4, 6 – master drive clutch; 3 – variable reducer; 5 – fly-wheel; 7 – engine; 8 – enter of electric line; 9 – exit of electric line; 10 – disk (wheel) of rotor (rigid attachment to shaft 17); 11 – motionless conductor (rigid attachment to stator); 12 – electric current; 13 – sliding contact; 14, 15 – exit conductor; 16 –

double switch from electric line 14 to conductor 11; 18 – sliding contact; 19 – mercury; 20 – electric ring; 21 – thin film; 22 – electric rail.

The sketch of the variable reducer is shown in fig. 12-3. The tape (inertial transfer roll) 3 rotates from shaft 1 (electric generator) to shaft 2 (fly-wheel). In starting position the tape (roll). diameter d_1 of shaft 1 is big while the tape (roll) diameter d_2 of the fly-wheel is small and rotation speed of electric generator is small. During the rotation, the tape (roll) diameter of shaft 1 decreases, while the corresponding diameter around shaft 2 increases and the rotation speed of the electric generator increases (assuming a correct design of the reducer). The total change of the rotation speed is $(d_1/d_2)^2$. For example, if $d_1/d_2 = 7$, the total change of rotary speed is 49. This way the rotation speed of the electric generator either increases or stays constant in spite of the fact that the rotary speed of the flywheel is decreasing. The multi-stage electric generator achieves the additional increasing of voltage. Its' sections turn on in series.

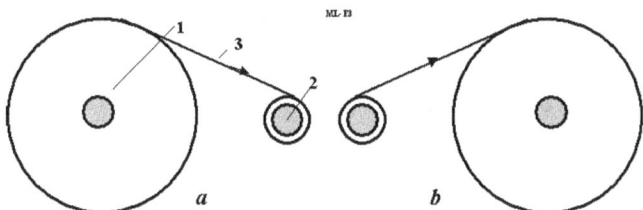

Fig. 12-3. Variable reducer. (*a*) Start position; (*b*) final position. *Notations*: 1 – shift of electric generator; 2 – shift of fly-wheel; 3 – tape (inertial transfer roll).

Conclusion

The research shows the magnetic launcher can be built by the current technology. This significantly (by a thousand times) decreases the cost of space launches. Unfortunately, if we want to use the short rail way (412 m), any launcher request a big acceleration about $7.5 \cdot 10^3 g$ and may be used only for unmanned, hardened payload. If we want design the manned launcher the rail way must be 1100 km for acceleration $a = 3g$ (untrained passengers) and about 500 km ($a = 6g$) for trained cosmonauts.

Our design is not optimal. For example, the computation shows, if we increase our rail track only by 15 m, we do not need gas gun initial acceleration. That significantly decreases the cost of installation and simplifies its construction.

The reader can recalculate the installation for his own scenarios.

13. Lower Current and Plasma Magnetic Railguns*

It is well-known that the magnetic railgun theoretically allows a very high 'exhaust velocity' of projectile. The USA and England have tried to research and develop working railgun installations. However the researchers had considerable problems in testing. The railgun requests very high (millions of amperes,) electric current (but low voltage). As result the rails and contacts burn and melt. The railgun can make only ONE shot between repairs, cannot shoot a big and high speed projectile, and has low efficiency.

The heat and inductive losses of railgun depend upon the square of electric current. If we decrease electric current by ten times, we decrease the losses by one hundred times. But the current design of railgun does not allow decreasing the current because that leads to loss --also by the square-- of electromotive force.

In this article the author describes new ideas, theory and computations for design of new magnetic lower cost accelerators for railgun projectiles and space apparatus. This design decreases the requested electric current (and loss) in hundreds times. This design requires a similar increase in voltage (because the energy for acceleration is the same). But no super heating, burn and melting rails, contacts, or big losses. The power and mass of projectiles and space apparatus can be increased in a lot of times. High voltage current does not require special low voltage equipment and may be used directly from the electric stations, saving huge amounts of money.

Author also suggests a new plasma magnetic accelerator, which has no traditional sliding mechanical contacts, significantly decreases the mass of electrical contacts and increases the useful mass of projectile in comparison with a conventional railgun.

Important advantages of the offered design are the lower (up to some tens of times) usage of electric current of high voltage and a very high efficiency coefficient closeto 95% (compare with efficiency of the current railgun which equals 20 – 40%). The suggested accelerators may be produced by present technology.

The projects of railguns are computed herein.

----

*See also [4], http://www.scribd.com/doc/24057930 . http://www.scribd.com/doc/31090728
http://www.archive.org/details/Macro-projectsEnvironmentsAndTechnologies

Description of Innovations and Problems in AB-Launchers

Low current multi-loop Railguns

Description of multi-loop Launchers . The conventional magnetic accelerator (railgun) is shown in fig. 13-1. That contains two the conductive rails connected by a **sliding** jumper. Electric current produces the magnetic field and magnetic force. The jumper accepts the magnetic force and accelerates the projectile. Main defects of conventional rail gun: The rail gun requires a gigantic current (millions of amperes) of low voltage, rails have large electric resistance, strongly heating up, contacts burn, installation is damaged and requires repair after every shot. The energy charge is high (small coefficient of efficiency). You see the gigantic plasma column behind the small projectile in fig.13-2 (left)). The repulsive force between rails is gigantic (thousands of tons) and installation is thus heavy and expensive if it is to survive a single shot.

Description.. The fig. 13-3a shows the principal scheme of the conventional railgun. The installation of fig.3a includes the long vertical wire 2, moved jumper 8, sliding contacts 7 and electric source 6. The electric current produces the magnetic field 3 (magnetic column), the magnetic field creates the vertical 4 and horizontal 5 forces. Vertical force 4 accelerates the useful load at top of the installation.

This design is used in a rail gun [4] but the author made many innovations that allow applying these ideas to this new application as the efficient magnetic accelerators. Some of them are listed below.

Innovations. The figs. 13-1b-1f show schemes of the suggested accelerators The author offers the following innovations having the next advantages:

1) *Version 1* (fig. 13-1b). The horizontal wire (fig. 13-1a, former sliding jumper 8 of fig. 13-1a) is made in a form of closed-loop spool (fig. 13-1b, #10). The lower part of this spool (fig. 13-1b, #11) located in place of former jumper near the magnetic field of vertical wire (fig. 13-1b). The top part of this spool (fig. 13-1b, #10) located at top – out of the magnetic field of vertical wire 2. As the result the magnetic field of the vertical wire 2 activates only on horizontal wires 11. But now we have here not one wire with current i, we have n wires and current ni. The magnetic force 4 and requested 4 voltage increases by n times! The force spool 10 can have some hundreds loops and force will be more in same times thanin case of fig. 13-1a. For same vertical force the electric current in the vertical connection wires 2 may be decreases in $n_1 = n$ times, where n_1 is number of horizontal loops. The electric current may be relatively small (only some tens of thousands of amperes, not millions) and of the high voltage. The request vertical wire may be relatively thin. The heating, contact and inductive losses decrease in n_1 times, where n_1 is number of horizontal loops. The number of contacts $N=2$ is same (not increases).

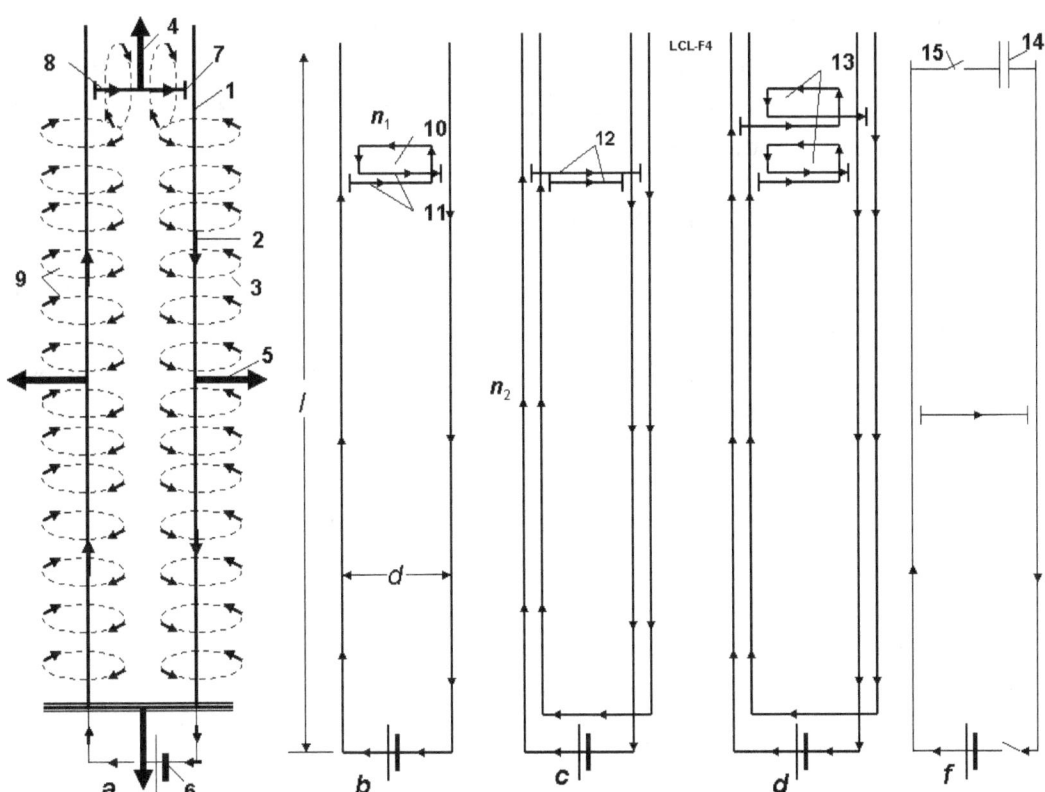

Fig. 13-1. Conventional and low current launchers. (*a*) Conventional high current and low voltage railgun; (*b*) Offered low current launcher with the wire horizontal multi-loops (version 1); (*c*) Offered low current launcher with the wire vertical multi-loops (version 2); (*d*) Offered low current launcher with the wire horizontal and vertical muiti-loops (version 3); (*f*) RailGun with condenser. (version 4). *Notations*: 1 – installation, 2 – vertical wire, 3 – magnetic field from vertical wire; 4 – moved vertical force from jumper; 5 – magnetic force from vertical electric wire, 6 – electric source, 7- sliding contact, 8 – horizontal jumper, 9 - magnetic column, 10 – multi-loop spool, 11, 12 – horizontal force wire connected in one bunch, 13 – force multi-loops spools connected in one spool, 14 – condenser, 15 – electric switch.

2) *Version 2* (fig.13-1c). Installation contains the vertical closed loops n_2. For same vertical force the electric current decreases in n_2 times, the voltage increases by n_2 times. The number of contacts N also increases by n_2 times. But heating of every contact decreases by n_2^2. Common electric heat, contact and inductive losses decrease by n_2 times.

3) *Version 3* (fig. 13-1d). That is composition of versions 1 and 2. For same vertical force the electric current decreases by $n_1 n_2$ times. The number of contacts N increases in n_2 times. But heating of every contact decreases by $n_1 n_2$. Common electric heat, contact and inductive losses decrease in $n_1 n_2$ times.

4) *Version 4* (fig.13-1f). The main electric loss in conventional railgun is an inductive loss, which produces a gigantic inductive current and plasma flash. This loss may be significantly decreases by switching the condenser in the end of the projectile track.

The main innovation is a top loop 10 (right angle spool [4]), which increases the number of horizontal wires 11 (multi-loops), magnetic intensity in area under 11 and lift force 4. We can make a lot of loops up to some hundreds and increase the lift force by hundreds of times. For a given lift force we can decrease the required current in many times and decreases the mass of the source wire 2. That does not necessarily mean that we decrease the required electric energy (power) because the new installation needs a higher voltage.

Multi-loop Railguns with permanent magnets

Main function of the vertical wire 2 (fig. 13-1a) creates the magnetic field between wire 2. This field interacts with the magnetic field from a jumper 8 (fig. 13-1a) or the horizontal wires 11 (fig. 13-1b), 12 (fig. 13-1c) and creates the vertical force 4 (fig. 13-1a). For getting enough force requires a high electric current. However, the needed intensity magnetic field we can produce by means of conventional magnets.

This idea explored in multi-loop launcher (accelerator) is offered in fig. 13-2. The launcher has two strong magnets 2 and the multi-loop spool 6. The spool connects through the sliding contacts 5 (fig. 13-2a) to the electric source 3. The design of fig. 13-2c has the spring wires 9 and not has the sliding contacts 5. The suggested launcher has two the motive magnetic jumpers 7, which significantly increase and close the magnetic lines in the lower part 8 of the force spool 6. The top part of the force spool has not the magnetic jumper 7 and does not produce the opposing motive force.

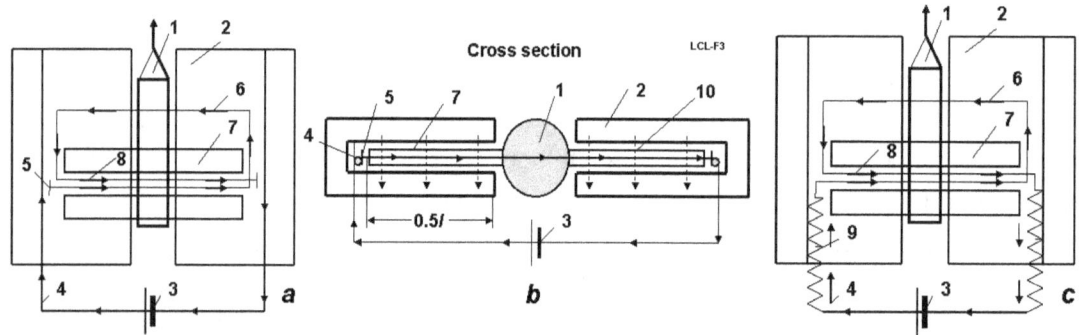

Fig. 13-2. Offered launcher with the permanent magnets. (*a*) Top view; (*b*) Cross section; (*c*) Launcher having the spring wires. *Notations*: 1 – projectile, 2 – magnet, 3 – electric source, 4 – electric current, 5 – sliding

contact, 6 – force (jumper) multi-loop spool (top part); 7 - magnetic jumper; 8 – lower part of the multi-loop top (force) electric spool; 9 – the spring electric wire; 10 – magnetic lines; 0.5*l* is length of a magnetic active zone in one side of wire.

The offered accelerator is closest to the direct current linear electric engine but has four differing important features: It has a spool part of it located out of the strong magnetic lines, the installation has the motive magnetic jumpers, the installation uses constant non-interrupted current, launcher can omit the sliding contacts (fig. 13-4c).

This magnetic accelerator may be suitable for a space launcher having small accelerations.

Plasma magnetic launcher

The jumper 4 (fig. 13-1a), force spools 10, 12, 13 (figs. 13-1b,c,d), magnetic jumpers 7 (fig. 13-2) can have a big mass and significantly decrease the useful load (in 2 – 3 and more times). For decreasing of this imperfect the author offers the plasma jumper (fig. 13-3). Plasma in jumper has very small mass because plasma has a small density. The plasma can have a high conductivity closed to a metal conductor. But plasma conductor (jumper) can have a big cross-section area and a low electric resistance. Electric conductivity of plasma does not depend from its density. The plasma may be very rarefied. That means the head transfer to walls of a channel may be very small and so not damage them. For example the plasma of Earth radiation belts has a million degrees but spaceman and space apparatus not damage.

The thin wire may be initial initiator of a of plasma cable. Then a plasma conductor supports the big current.

The second advantage of plasma launcher is a gas sliding contact. That cannot burn and is more robust against damage.

The offered plasma cable may be used in other technical fields [2] –[4]. This problem needs further research.

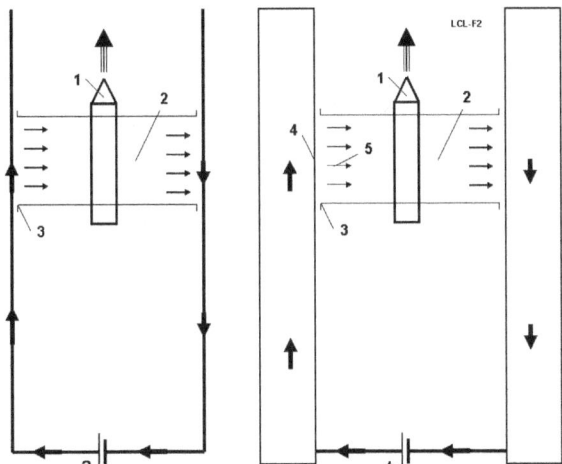

Fig.13-3. Offered plasma launchers. (***a***) Launcher having conventional vertical wire and plasma jumper; (***b***) Launcher having vertical and horizontal (jumper) plasma wires. *Notations:* 1- projectile, 2 – horizontal plasma jumper, 3 – gas sealing, 4 – conducting tube, 5 – electric current into plasma.

Conclusion

In this article the author describes the new ideas, theory and computations for design the new low electric current launchers for the railgun projectile and space apparatus.

Important advantage of the offered design is the lower (up some tens times) used electric current of high voltage and the very high inductive efficiency coefficient close to 0.9 (compared with efficiency of the current railgun equal to 20 – 40%). The suggested launchers may be produced by present technology.

The problems of needed electric energy become far simpler. At first, AB-Launchers have a high efficiency and spend in 2 – 3 times less energy than a conventional railgun; the second, AB-Launcher uses the high voltage energy closed to a voltage of the electric stations. That means the power electric station can be directly connected to AB-Launcher in period of acceleration without expensive transformers and condensers. The power of strong electric plant is enough for launching the space apparatus of some hundreds of kilograms.

The offered magnetic space launcher is a thousand times cheaper than the well-known cable space elevator. NASA is spending for research of space elevator hundreds of millions of dollars. A small part of this sum is enough for R&D of the magnetic launcher and to make a working model.

The proposed innovation (milti- electric AB-spool, permanent magnetic rails, plasma magnetic launcher) allows also solving the problem of the conventional railgun (having the projectile speed 3 -5 km/s). The current conventional railgun uses a very high ampere electric current (millions A) and low voltage. As the result the rails corrode, burn, melt The suggested AB-spool allows decreases the required the electric current by tens of times (simultaneously the required voltage is increased by the same factor).
Small cheap magnetic prototypes would be easily tested.

The computed projects are not optimal. That is only illustration of an estimation method. The reader can recalculate the AB-Launchers for his own scenarios (see also [1]-[4]).

14. Superconductivity Space Accelerator *

In this Chapter the author describes a new idea, theory and computations for design of a new magnetic low cost accelerator for railgun projectile and space apparatus. The suggested design does not have the traditional current rails and sliding contacts. This accelerates the projectile and space apparatus by a magnetic column which can have a length of some kilometers, produces a very high acceleration and projectile (apparatus) final speed of up to 8 - 10 km/s.
Important advantages of the offered design is the low (up to some thousands of times) used electric current of high voltage and very high efficiency coefficient close to 1 (compare with efficiency of the current railgun which equals 20 – 40%). The suggested accelerator may be produced by present technology.

The projects: railgun and space accelerator are computed in [4].
----
* Magnetic Space AB-Accelerator. http://www.scribd.com/doc/26885058 , or [4].

Description of Innovations and Problems

New type of magnetic acceleration (magnetic AB-column)

1. Description of Innovations. The conventional magnetic accelerator (railgun) is shown in fig.1. That contains two the conductive rails connected by a **sliding** jumper. Electric currents produce the magnetic field and magnetic force. The jumper accepts the magnetic force and accelerates the projectile. Main defects of conventional rail gun: The rail gun requires a gigantic current (millions of amperes) of low voltage, rails have large electric resistance, strongly heating up, contacts burn, installation is damaged and requires repair after every shot. The energy charge is high (small coefficient of efficiency. You see the gigantic plasma column behind the small projectile in [4] Ch.6, fig.2). The repulsive force between rails is gigantic (thousands of tons) and installation is thus heavy and expensive if it is to survive a single shot.

Description. The fig.1a shows the suggested accelerator without the force and connection spools. Installation includes the long vertical loop from electric wire 1 and electric source 6. The electric current 2 produces the magnetic field 3 (magnetic column), the magnetic field creates the vertical 4 and horizontal 5 forces. These forces balance the film (or filament, fiber) connection 15. Vertical force 4 accelerates the useful load at top of the installation and supports wires and film connection).

This design is used in a rail gun [4] but the author made many innovations that allow applying this idea to this new application as an efficient magnetic accelerator. Some of them are listed below.

Innovations. The author offers the following innovations having the next advantages:

1) The horizontal wire (fig. 14-1a, #2a, former sliding jumper) is made in a form of closed-loop spool (fig. 14-1b, #11). The lower part of this spool (fig. 14-1b, #12) located in place of former jumper near the magnetic field of vertical wire (fig. 14-1b). The top part of this spool (fig. 14-1b, #13) located at top – out of the magnetic field of vertical wire 2. As the result the magnetic field of the vertical wire 2 activates only on horizontal wires 12. But now we have here not one wire 2a with current *i*, we have *n* wires and current *ni*. The magnetic force 17 and requested voltage increases by *n* times! The force spool 11 can have some thousands loops and force 17 will be more in same times then in case 2a of fig. 14- 3a. That means the electric current in the connection wires 2 may be relatively small (only some thousands of amperes, not millions) and of high voltage. The request vertical wire may be relatively thin.

2) Application of special connection spool. The accelerator has the connection spool 7 (fig. 14-1, detail spool in fig. 14-2). That allows increases the acceleration distance up to some kilometers and deletes the sliding electric contacts.

3) The wire is superconductive and has special design (fig. 14-2). That allows simply cooling the wires in a short time (the superconductivity needs a low temperature). Their cooling time may be short because requested time of acceleration may be short. For example, rail gun shot lasts about 0.1 second, the space apparatus acceleration is about tens seconds.

4) Absence of long rails. 'Confinement recoil' of projectile is accepted by the magnetic columns 3a (fig. 14-1a) from vertical wires [4].

The main innovation is a top loop 11 (right angle spool [4]), which increases the number of horizontal wires 12, magnetic intensity in area 17 and lift force 16. We can make a lot of loops up to some thousands and increase the lift force by thousands of times. For a given lift force we can decrease the required current in many times and decreases the mass of source wire 2. That does not necessarily mean that we decrease the required electric power because the new installation needs a higher voltage. The proposed construction creates the MAGNETIC COLUMN 3a that produces a lift force some thousands of times more than a conventional rail gun.

The second important innovation is the connection spool, which increases the acceleration distance, deletes the sliding contacts and heavy rails.

Quadratic magnetic column. The quadratic four wire magnetic column ([4], Fig.2) is more efficient, stable, safe, reliable and controlled than two wire magnetic column Fig. 14-1. It can control of space ship direction (by changing current in vertical wires). It is important for high altitude space apparatus.

2. Connection spools. The connection spools can be located at accelerator and at the ground. Every design has its advantages and limitations. Three constructions of the connection spools are presented in fig. 14-2. The first design (fig. 14-4a) has immobile vertical spools. That is simple but results in a limited safe high speed of the launched apparatus and requires the connection device 9. The second design (fig. 14-2b) has the horizontal spool rotated by an engine. This connection spool is limited by the safe rotary speed of the spool and also requires the connection device 9. The third design (fig. 14-2c) has engine and also limited by a safety rotary of spool, but it does not request the connection device 9 because the wires are connected by fiber before spooling in the connection spool.

The limitation is about 1 – 3 km/s for the current artificial fiber (whiskers) having a safe tensile stresses about 200 - 1000 kgf/mm^2. But if we use nanotubes, the limit is more by 5 -10 times.

3. Cooling system of superconductive wire. The current superconductive conductors do not spend electric energy and pass very large electric current densities, but they require an cooling system because the current superconductive materials have the critical temperature of about 100 - 180 K (see Table #1 below). The wire located into Earth's atmosphere (up 50 – 80 km) needs cooling.

However, the present computational methods of heat transfer are well developed and the weight and the induced expenses for cooling are small (for example, cooling by liquid nitrogen) [4] (see also Computation and Projects sections).

The suggested design of a cooled superconductive wire is present in fig. 14-3. The wire contains two elastic tubes. The insulated internal tube is coated inside-- the superconductive layer, outside is coated--the highly reflective layer. The outer tube is made from the strong artificial fiber and covered by the highly reflective layers. The space between tubes is vacuumed or filed by air (it is worse but may apply for short cooling time) or heat protection.

The wire works the following way. The liquid nitrogen (77 K) from special heat protection capsule is injected into the internal tube in many places (needles) and instantly cooling the superconductive layer to lower than the critical temperature.

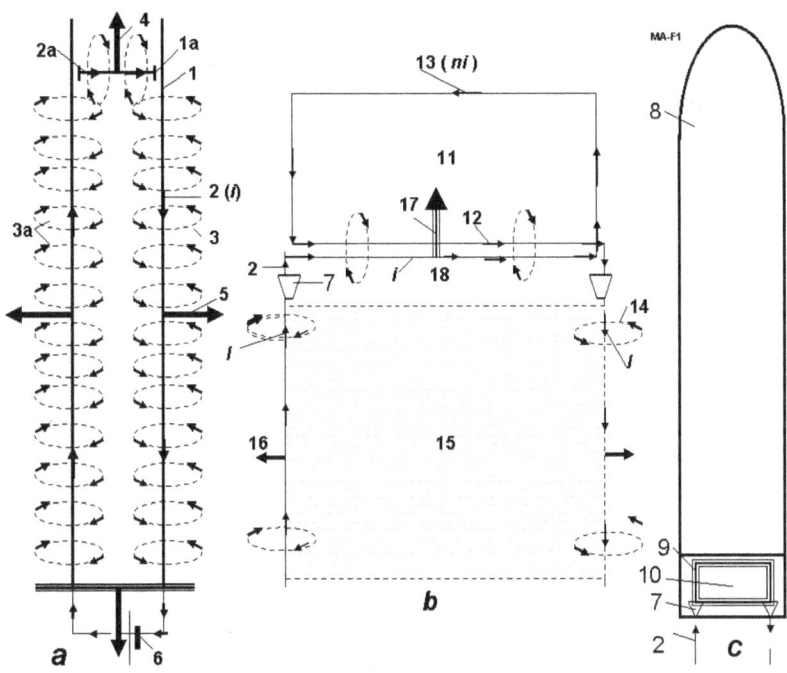

Fig. 14-1. Principal sketch of the Magnetic AB-Accelerator. *Notations*: (**a**) Principal sketch of conventional (one turn) magnetic accelerator, (**b**) – Multi-loop force spool at top of accelerator (same force spool located at ground), (**c**) – rocket (projectile) with detachable magnetic accelerator at bottom. *Parts*: 1 – magnetic installation with magnetic column, 1a – sliding contct, 2 – vertical wire and direction of electric current *i*, 2a – horizontal wire (jumper) and direction of electric current *i*, 3 – magnetic field from vertical electric wire, 3a – magnetic column, 4 – magnetic force from horizontal electric wire (jumper), 5 – magnetic force from vertical electric wire, 6 – electric source, 7 – connection spool, 8 – rocket (projectile), 9 – force spool, 10 – magnetic accelerator, 11 – wire multi-loop superconductive force spool at top (same spool located also on Earth surface), 12 – lower wires of loop (force spool), 13 – top wires of loop (force spool), 14 – magnetic field from vertical wire, 15 – thin film or artificial fiber connected the vertical wires for compensation the magnetic repulse force, 16 – repulsive magnetic force, 17 – acceleration force, 18 – area strong magnetic field, *i* – electric current in vertical wire, *ni* – electric current in the force spool.

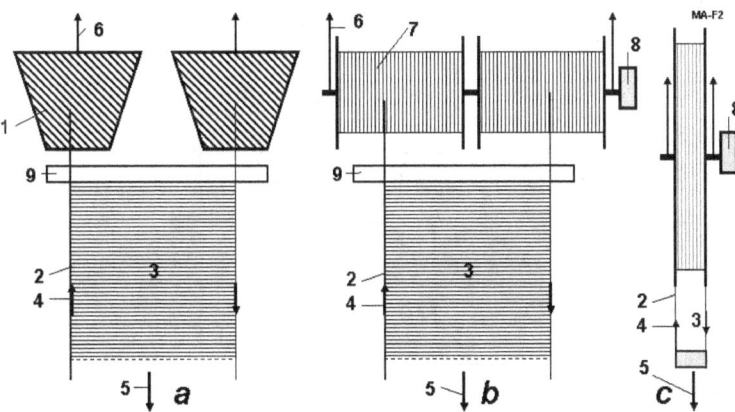

Fig. 14-2. Possible installations of connection spools for AB-Accelerator. *Notations*: (***a***) Immobile vertical connection spool, (***b***) Rotated horizontal connection spool, (***c***) Type connection spool. Parts: 1 – immobile

vertical connection spool, 2 – vertical superconductive wire, 3 – filaments connected the vertical wire and keeping the repulsive magnetic force of vertical wires, 4 – electric current, 5 – motion of connection wire, 6 – vertical wire to the force spool, 7 – connection wire in connection spool, 8 – engine for rotation of the connection spool, 9 – device for connection the vertical wires by filaments.

While the nitrogen evaporates the temperature is 77 K and installation can accelerate the projectile or space apparatus. After acceleration the accelerator separates, wires are spooling and installation is ready for next shot.

4. Superconductive materials.

There are hundreds of new superconductive materials (type 2) having critical temperature 70 ÷ 120 K and more. Some of the superconductable materials are presented in Table 1 (2001). The widely used $YBa_2Cu_3O_7$ has mass density 7 g/cm^3.

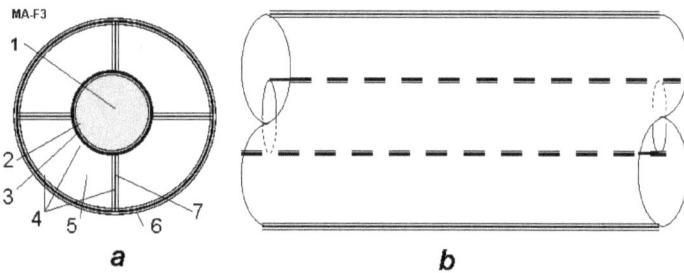

Fig. 14-3. Superconductive wires. (**a**) Cross-section of superconductive wires, (**b**) – side view. *Notations*: 1 - strong elastic tube (internal part is used for cooling of superconductive layer by liquid nitrogen, external part is used for reflective layer), 2 – superconductive layer, 3 - insulator, 4 – high reflective layer, 5 – vacuum or air, or heat-insulated material (fiber), 6 – strong outer tube (internal and external surface is covered by reflective coating), 7 – connection the internal and outer tubes.

Table 1. Transition temperature T_c and upper critical magnetic field $B = H_{c2}(0)$ of some examined superconductors AIP, Physics desk references, 3rd ed., p. 752.

Crystal	T_c (K)	H_{c2} (T)
La$_{2-x}$Sr$_x$CuO$_4$	38	≥80
YBa$_2$Cu$_3$O$_7$	92	≥150
Bi$_2$Sr$_2$Ca$_2$Cu$_3$O$_{10}$	110	≥250
TlBa$_2$Ca$_2$Cu$_3$O$_9$	110	≥100
Tl$_2$Ba$_2$Ca$_2$Cu$_3$O$_{10}$	125	≥150
HgBa$_2$Ca$_2$Cu$_3$O$_8$	133	≥150

The last decisions are: Critical temperature is 176 K, up to 183 K. Nanotube has critical temperature of 12 - 15 K,

Some organic matters have a temperature of up to 15 K. Polypropylene, for example, is normally an insulator. In 1985, however, researchers at the Russian Academy of Sciences discovered that as an oxidized thin-film, polypropylene have a conductivity 10^5 to 10^6 that is higher than the best refined metals.

Boiling temperature of liquid nitrogen is 77.3 K, air 81 K, oxygen 90.2 K, hydrogen 20.4 K, helium 4.2 K. Specific density of liquid air is 920 kg/m^3, nitrogen 804 kg/m^3; evaporation heat is liquid air is 213 kJ/kg, nitrogen 199 kJ/kg.

Unfortunately, most superconductive material is not strong and needs a strong covering for structural support.

5. Advantages. The offered magnetic accelerator has big advantages in comparison with railguns and other space launchers. Compare it with the space rocket.

1. The AB-Accelerator is very cheap. The cost is about one million USD (rail gun) to some millions (space launcher) [4].
2. The consumables cost is very small and primarily the needed electric energy (about 3 – 5 $/kg)[4].
3. The productivity is very high (tens launches in day).
4. The accelerator uses the current well developed technology and may be researched and developed in a short time.
5. The accelerator (special platform) may initially to accelerate current rockets up to speed 700 – 1000 m/s and lift them to the altitude 5 – 10 km. That increases the payload the current rockets up 50% .
6. Accelerator uses the high voltage (up to 1 MV) electric currency. That allows to directly connect accelerator to current power electric stations (in night time during periods of slack power use) and to launch a space apparatus without expensive electric energy storage in the form of capacitors (used in present time).
7. Accelerator has a coefficient of energy efficiency closed to 1. It is the most efficient among the known space launchers.

In comparison with current Railgun the suggested accelerator has the following advantages:

1. No problem with burn of rail and contact.
2. No damage and repair of installation after every shot.
3. Limit in speed of projectile is high (7 – 9 km/s).
4. No limit to mass of projectile.
5. Installation is cheaper.
6. Installation requires ~2 – 3 times less of energy than same output conventional railgun.

6. Application and further development. Idea of the magnetic AB-column may be applied to the suspending of houses, buildings, towns, multi-floor cities, to a small flying city-state located over ocean in the international water, (avoiding some of the liabilities of sea-surface communities during storms) to levitating space stations, to communication masts and towers [4]. This idea may be easily tested in small cheap magnetic constructions for simple projects, on a small scale.

Conclusion

In this chapter the author describes new idea, theory and computations for design the new magnetic low cost accelerator for railgun projectile and space apparatus. The suggested design does not have the traditional current rails and sliding contacts. That accelerates the projectile and space apparatus by a magnetic column which can have a length of some kilometers, produces the very high acceleration and the projectile (apparatus) speed up 8 km/s.

Important advantage of the offered design is the lower (up some thousands times) used electric current of high voltage and the very high efficiency coefficient close to 1 (compared with efficiency of the current railgun equal to 20 – 40%). The suggested accelerator may be produced by present technology.

The important advantage of the offered method for space apparatus is following: The method does not need designing new rockets. What is needed is to design only a simple accelerator (accelerate platform). Any current rocket may be installed on this platform and accelerated up high speed and lifted on high altitude before started. That radically increases payload and decreases the cost of launching. The platform (force spool) and wires disconnects from the rocket after acceleration. Platform returns by parachute, the wires reel back to start.

The problems of needed electric energy become far simpler. At first, AB-Accelerator has very high efficiency and spend in 2 – 3 times less energy then a conventional railgun; the second, AB-Accelerator uses the high voltage energy same with voltage the electric stations. That means the power electric station can be directly connected to AB-Accelerator in period of acceleration without expensive transformers and condensers. The power of strong electric plant is enough for launching the rocket (space apparatus) of some hundreds of tons.

The offered magnetic space accelerator is a thousand times cheaper than the well-known cable space elevator. NASA is spending for research of space elevator hundreds of millions of dollars. A small part of this sum is enough for R&D of the magnetic accelerator and make a working model!

The proposed innovation (upper electric AB-spool) allows also solving the problem of the conventional railgun (having the projectile speed 3 -5 km/s). The current conventional railgun uses a very high ampere electric current (millions A) and low voltage. As the result the rails burn. The suggested superconductive AB-spool allows decreases the required electric current by thousands of times (simultaneously the required voltage is increased by the same factor). No rails, therefore no damage to the rails.
Small cheap magnetic prototypes would be easily tested.

The computed projects (in [4]) are not optimal. That is only illustration of an estimation method. The reader can recalculate the AB-Accelerator for his own scenarios (see also [1]-[4]).

15. Converting of Matter to Nuclear Energy by AB-Generator and Photon Rocket*

Author offers a new nuclear generator which allows to convert any matter to nuclear energy in accordance with the Einstein equation $E=mc^2$. The method is based upon tapping the energy potential of a Micro Black Hole (MBH) and the Hawking radiation created by this MBH. As is well-known, the vacuum continuously produces virtual pairs of particles and antiparticles, in particular, the photons and anti-photons. The MBH event horizon allows separating them. Anti-photons can be moved to the MBH and be annihilated; decreasing the mass of the MBH, the resulting photons leave the MBH neighborhood as Hawking radiation. The offered nuclear generator (named by author as AB-Generator) utilizes the Hawking radiation and injects the matter into MBH and keeps MBH in a stable state with near-constant mass.

The AB-Generator can produce gigantic energy outputs and should be cheaper than a conventional electric station by a factor of hundreds of times. One also may be used in aerospace as a photon rocket or as a power source for many vehicles.

Many scientists expect the Large Hadron Collider at CERN will produce one MBH every second. A technology to capture them may follow; than they may be used for the AB-Generator.

----

* Presented as Paper AIAA-2009-5342 in 45 Joint Propulsion Conferences, 2–5 August, 2009, Denver, CO, USA. See also Converting of Any Matter to Nuclear Energy by-AB-Generator
Converting of any Matter to Nuclear Energy by AB-Generator and Aerospace ,
http://www.archive.org/details/ConvertingOfAnyMatterToNuclearEnergyByAb-generatorAndAerospace ,

AB-Generator of Nuclear Energy and some Innovations

Simplified explanation of MBH radiation and work of AB-Generator ([7], Fig.5). As known, the vacuum continuously produces, virtual pairs of particles and antiparticles, in particular, photons and anti-photons. In conventional space they exist only for a very short time, then annihilate and return back to nothingness. The MBH event horizon, having very strong super-gravity, allows separation of the particles and anti particles, in particular, photons and anti-photons. Part of the anti-photons move into the MBH and annihilate with photons decreasing the mass of the MBH and return back a borrow energy to vacuum. The free photons leave from the MBH neighborhood as Hawking radiation. That way the MBH converts any conventional matter to Hawking radiation which may be converted to heat or electric energy by the AB- Generator. This AB- Generator utilizes the produced Hawking radiation and injects the matter into the MBH while maintaining the MBH in stable suspended state.

Note: The photon does NOT have rest mass. Therefore a photon can leave the MBH's neighborhood (if it is located beyond the event horizon). All other particles having a rest mass and speed less than light speed *cannot* leave the Black Hole. They cannot achieve light speed because their mass at light speed equals infinity and requests infinite energy for its' escape—an impossibility.

Description of AB- Generator. The offered nuclear energy AB- Generator is shown in fig. 15-1. That includes the Micro Black Hole (MBH) 1 suspended within a spherical radiation reflector and heater 5. The MBH is supported (and controlled) at the center of sphere by a fuel (plasma, proton, electron, matter) gun 7. This AB-Generator also contains the 9 – heat engine (for example, gas, vapor turbine), 10 – electric generator, 11 – coolant (heat transfer agent), an outer electric line 12, internal electric generator (5 as antenna) with customer 14.

Work. The generator works the following way. MBH, by selective directional input of matter, is levitated in captivity and produces radiation energy 4. That radiation heats the spherical reflector-heater 5. The coolant (heat transfer agent) 11 delivers the heat to a heat machine 9 (for example, gas, vapor turbine). The heat machine rotates an electric generator 10 that produces the electricity to the outer electric line 12. Part of MBH radiation may accept by sphere 5 (as antenna) in form of electricity.

The control fuel guns inject the matter into MBH and do not allow bursting of the MBH. This action also supports the MBH in isolation, suspended from dangerous contact with conventional matter. They also control the MBH size and the energy output.

Any matter may be used as the fuel, for example, accelerated plasma, ions, protons, electrons, micro particles, etc. The MBH may be charged and rotated. In this case the MBH may has an additional suspension by control charges located at the ends of fuel guns or (in case of the rotating charged MBH) may have an additional suspension by the control electric magnets located on the ends of fuel guns or at points along the reflector-heater sphere.

Innovations, features, advantages and same research results

Some problems and solutions offered by the author include the following:

1) A practical (the MBH being obtained and levitated, details of which are beyond the scope of this paper) method and installation for converting any conventional matter to energy in accordance with Einstein's equation $E = mc^2$.

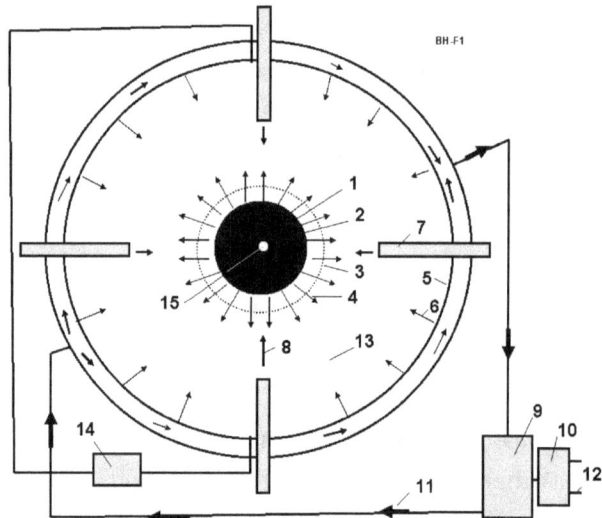

Fig. 15-1. Offered **nuclear-vacuum energy AB- Generator**. *Notations*: 1- Micro Black Hole (MBH), 2 - event horizon (Schwarzschild radius), 3 - photon sphere, 4 – black hole radiation, 5 – radiation reflector, antenna and heater (cover sphere), 6 – back (reflected) radiation from radiation reflector 5, 7 – fuel (plasma, protons, electrons, ions, matter) gun (focusing accelerator), 8 – matter injected to MBH (fuel for Micro Black hole), 9 – heat engine (for example, gas, vapor turbine), 10 – electric generator connected to heat engine 9, 11 – coolant (heat transfer agent to the heat machine 9), 12 – electric line, 13 – internal vacuum, 14 – customer of electricity from antenna 5, 15 – singularity.

2) MBHs may produce gigantic energy and this energy is in the form of dangerous gamma radiation. The author shows how this dangerous gamma radiation Doppler shifts when it moves against the MBH gravity and converts to safely tapped short radio waves.

3) The MBH of marginal mass has a tendency to explode (through quantum evaporation, very quickly radiating its mass in energy). The AB- Generator automatically injects metered amounts of matter into the MBH and keeps the MGH in a stable state or grows the MBH to a needed size, or decreases that size, or temporarily turns off the AB- Generator (decreases the MBH to a Planck Black Hole).

4) Author shows the radiation flux exposure of AB- Generator (as result of MBH exposure) is not dangerous because the generator cover sphere has a vacuum, and the MBH gravity gradient decreases the radiation energy.

5) The MBH may be supported in a levitated (non-contact) state by generator fuel injectors.

AB-Generator as Photon Rocket

The offered AB- Generator may be used as the most efficient photon propulsion system (photon rocket). The photon rocket is the dream of all astronauts and space engineers, a unique vehicle) which would make practical interstellar travel. But a functioning photon rocket would require gigantic energy. The AB- Generator can convert any matter in energy (radiation) and gives the maximum theoretical efficiency.

The some possible photon propulsion system used the AB –Generator is shown in Fig.15-2. In simplest version (*a*) the cover of AB generator has window 3, the radiation goes out through window and produces the thrust. More complex version (*c*) has the parabolic reflector, which sends all radiation in one direction and increases the efficiency. If an insert in the AB- Generator covers the lens 6 which will focuses the radiation in a given direction, at the given point the temperature will be a billions degree (see Equation (2)) and AB-Generator may be used as a photon weapon.

The maximal thrust *T* of the photon engine having AB- Generator may be computed (estimated) by equation:

$$T = \dot{M}c, \quad \text{N}, \qquad (26)$$

For example, the AB-generator, which spends only 1 gram of matter per second, will produce a thrust 3×10^5 N or 30 tons.

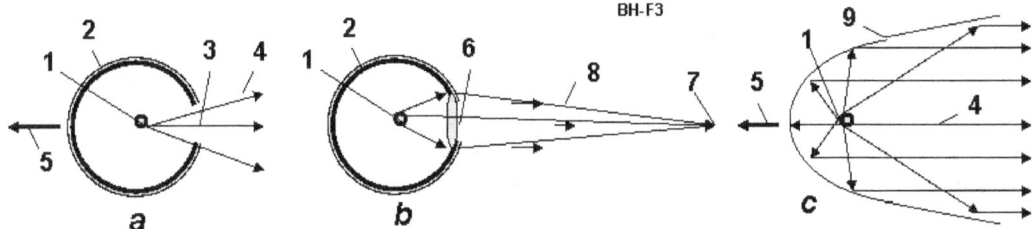

Fig. 15-2. AB-Generator as Photon Rocket and Radiation (Photon) Weapon. (*a*) AB- Generator as a Simplest Photon Rocket; (*b*) AB- Generator as focused Radiation (photon, light or laser) weapon; (*c*) Photon Rocket with Micro-Black Hole of AB-Generator. *Notations*: 1 – control MBH; 2 – spherical cover of AB-Generator; 3 – window in spherical cover; 4 – radiation of BH; 5 – thrust; 6 – lens in window of cover; 7 – aim; 8 - focused radiation; 9 – parabolic reflector.

Results:

1. Author has offered the method and installation for converting any conventional matter to energy according the Einstein's equation $E = mc^2$, where *m* is mass of matter, kg; $c = 3 \cdot 10^8$ is light speed, m/s.
2. The Micro Black Hole (MBH) is offered for this conversion.

3. Also is offered the control fuel guns and radiation reflector for explosion prevention of MBH.

4. Also is offered the control fuel guns and radiation reflector for the MBH control.

5. Also is offered the control fuel guns and radiation reflector for non-contact suspension (levitation) of the MBH.

6. For non contact levitation of MBH the author also offers:
   a) Controlled charging of MBH and of ends of the fuel guns.
   b) Control charging of rotating MBH and control of electric magnets located on the ends of

   the fuel guns or out of the reflector-heater sphere.

7. The author researches show the very important fact: A strong gamma radiation produced by Hawking radiation loses energy after passing through the very strong gravitational MBH field. The MBH radiation can reach the reflector-heater as the light or short-wave radio

   radiation. That is very important for safety of the operating crew of the AB- Generator.

8. The author researches show: The matter particles produced by the MBH cannot escape from MBH and can not influence the Hawking radiation.

9. The author researches show another very important fact: The MBH explosion (hundreds and

   thousands of TNT tons) in radiation form produces a small pressure on the reflector-heater (cover sphere) and does not destroys the AB-generator (in a correct design of AB-generator!). That is very important for safety of the operating crew of the AB-generator.

10. The author researches show another very important fact: the MBH cannot capture by oneself the surrounding matter and cannot automatically grow to consume the planet.

11. As the initial MBH can be used the Planck's (quantum) MBH which *may* be everywhere. The offered fuel gun may to grow them (or decrease them) to needed size or the initial MBH

    may be used the MBH produce Large Hadron Collider (LHC) at CERN. Some scientists assume LHC will produce one MBH every second (86,400 MBH in day). The cosmic radiation also produces about 100 MBH every year.

12. The spherical dome of MBH may convert part of the radiation energy to electricity.

13. A correct design of MBH generator does not produce the radioactive waste of environment.

14. The attempts of many astronomers find (detect) the MBH by a MBH exposure radiation will not be successful without knowing the following: The MBH radiation is small, may be detected only over a short distance, does not have specific frequency and has a variable long wavelength.

Discussing

We got our equations in assumption $\lambda/\lambda_o = r/r_o$. If $\lambda/\lambda_o = (r/r_o)^{0.5}$ or other relation, the all above equations may be easy modified.

The Hawking article was published 34 years ago (1974)[4]. After this time the hundreds of scientific works based in Hawking work appears. No facts are known which creates doubts in the possibility of Hawking radiation but it is not proven either. The Hawking radiation may not exist. The Large Hadron Collider has the main purpose to create the MBHs and detect the Hawking radiation.

Conclusion

The AB-Generator could create a revolution in many industries (electricity, car, ship, transportation, etc.). That allows designing photon rockets and flight to other star systems. The maximum possible efficiency is obtained and a full solution possible for the energy problem of humanity. These overwhelming prospects urge us to research and develop this achievement of science [4],[7].

16. Femtotechnology: the Strongest AB-Matter with Fantastic Properties and their Applications in Aerospace*

At present the term 'nanotechnology' is well known – in its' ideal form, the flawless and completely controlled design of conventional molecular matter from molecules or atoms. Such a power over nature would offer routine achievement of remarkable properties in conventional matter, and creation of metamaterials where the structure not the composition brings forth new powers of matter.

But even this yet unachieved goal is not the end of material science possibilities. The author herein offers the idea of design of new forms of nuclear matter from nucleons (neutrons, protons), electrons, and other nuclear particles. He shows this new 'AB-Matter' has extraordinary properties (for example, tensile strength, stiffness, hardness, critical temperature, superconductivity, supertransparency, zero friction, etc.), which are up to millions of times better than corresponding properties of conventional molecular matter. He shows concepts of design for aircraft, ships, transportation, thermonuclear reactors, constructions, and so on from nuclear matter. These vehicles will have unbelievable possibilities (e.g., invisibility, ghost-like penetration through any walls and armour, protection from nuclear bomb explosions and any radiation flux, etc.)

People may think this fantasy. But fifteen years ago most people and many scientists thought – nanotechnology is fantasy. Now many groups and industrial labs, even startups, spend hundreds of millions of dollars for development of nanotechnological-range products (precise chemistry, patterned atoms, catalysts, metamaterials, etc) and we have nanotubes (a new material which does not exist in Nature!) and other achievements beginning to come out of the pipeline in prospect. Nanotubes are stronger than steel by a hundred times—surely an amazement to a 19th Century observer if he could behold them.

Nanotechnology, in near term prospect, operates with objects (molecules and atoms) having the size in nanometer (10^{-9} m). The author here outlines perhaps more distant operations with objects (nuclei) having size in the femtometer range, (10^{-15} m, millions of times less smaller than the nanometer scale). The name of this new technology is femtotechnology.

----

*Femtotechnology: Design of the Strongest AB-Matter for Aerospace
http://www.archive.org/details/FemtotechnologyDesignOfTheStrongestAb-matterForAerospace   or
American Journal of Engineering and Applied Science, Vol. 2, #2, 2009, pp.501-514.
http://www.scipub.org/fulltext/ajeas/ajeas22501-514.pdf

Innovations and computations

Short information about atom and nuclei. Conventional matter consists of atoms and molecules. Molecules are collection of atoms. The atom contains a nucleus with proton(s) and usually neutrons (Except for Hydrogen-1) and electrons revolve around this nucleus. Every particle may be characterized by parameters as mass, charge, spin, electric dipole, magnetic moment, etc. There are four forces active between particles: strong interaction, weak interaction, electromagnetic charge (Coulomb) force and gravitational force. The nuclear force dominates at distances up to 2 fm (femto, 1 fm = 10^{-15} m). They are hundreds of times more powerful than the charge (Coulomb force and million-millions of times more than gravitational force. Charge (Coulomb) force is effective at distances over 2 fm. Gravitational force is significant near and into big masses (astronomical objects such as planets, stars, white dwarfs, neutron stars and black holes). Strong force is so overwhelmingly powerful that it forces together the positively charged protons, which would repel one from the other and fly apart without it. The strong force is key to the relationship between protons, neutrons and electrons. They can keep electrons into or near nuclei. Scientists conventionally take into attention only of the strong force when they consider the nuclear and near nuclear size range, for the other forces on that scale are negligible by comparison for most purposes.

Strong nuclear forces are anisotropic (non spherical, force distribution not the same in all directions equally), which means that they depend on the relative orientation of the nucleus.

Typical nuclear energy (force) is presented in fig. 16-1. When it is positive the nuclear force repels the other atomic particles (protons, neutrons, electrons). When nuclear energy is negative, it attracts them up to a distance of about 2 fm. The value r_0 usually is taken as radius of nucleus. Average interaction energy between to nucleus is about 8 MeV, distance where the attractive strong nuclear force activates is at about 1 – 1.2 fm.

2. AB-Matter. In conventional matter made of atoms and molecules the nucleons (protons, neutrons) are located in the nucleus, but the electrons rotate in orbits around nucleus in distance in millions times more than diameter of nucleus. Therefore, in essence, what we think of as solid matter contains a -- relatively! --'gigantic' vacuum (free space) where the matter (nuclei) occupies but a very small part of the available space. Despite this unearthly emptiness, when you compress this (normal, non-degenerate) matter the electrons located in their orbits repel atom from atom and resist any great increase of the matter's density. Thus it feels solid to the touch.

The form of matter containing and subsuming all the atom's particles into the nucleus is named *degenerate matter*. Degenerate matter found in white dwarfs, neutron stars and black holes. **Conventionally this matter in such large astronomical objects has a high temperature (as independent** particles!) and a high gravity adding a forcing, confining pressure in a very massive celestial objects. In nature, degenerate matter exists stably (as a big lump) to our knowledge only in large astronomical masses (include their surface where gravitation pressure is zero) and into big nuclei of conventional matter.

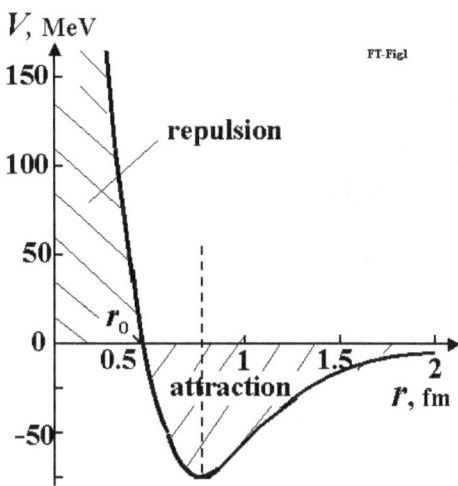

Fig. 16-1. Typical nuclear force of nucleus. When nucleon is at distance of less than 1.8 fm, it is attracted to nucleus. When nucleon is very close, it is repulsed from nucleus.
(Reference from http://www.physicum.narod.ru , Vol. 5 p. 670).

Our purpose is to design artificial small masses of synthetic degenerate matter in form of an extremely thin strong thread (fiber, filament, string), round bar (rod), tube, net (dense or non dense weave and mesh size) which can exist at Earth-normal temperatures and pressures. Note that such stabilized degenerate matter in small amounts does not exist in Nature as far as we know. Therefore I have named this matter **AB-Matter** (fig.16-2). Just as people now design by the thousands variants of artificial materials (for example, plastics) from usual matter, we soon (historically speaking) shall create many artificial, designer materials by nanotechnology (for example, nanotubes: SWNTs (amchair, zigzag, ahiral, graphen), MWNTs (fullorite, torus, nanobut), nanoribbon (plate), buckyballs (ball), fullerene. Sooner or later we may anticipate development of femtotechnology and create such AB-Matter. Some possible forms of AB-Matter are shown in fig.10. Offered technologies are шт [8]. The threads from AB-Matter are stronger by millions of times than normal materials. They can be inserted as reinforcements, into conventional materials, which serve as a matrix, and are thus strengthened by thousands of times (see computation section in [8]).

1. **Some offered technologies for producing** AB-Matter. One method of producing AB-Matter may use the technology reminiscent of computer chips ([8], fig.11). The stability and other method of production AB-matter considered in the article preparing for publication.

Various other means are under consideration for generation of AB-Matter, what is certain however that once the first small amounts have been achieved, larger and larger amounts will be produced with ever increasing ease. Consider for example, that once we have achieved the ability to make a solid AB-Matter film (a sliced plane through a solid block of AB-matter), and then developed the ability to place holes with precision through it one nucleon wide, a modified extrusion technique may produce AB-Matter strings (thin fiber), by passage of conventional matter in gas, liquid or solid state through the AB-Matter matrix (mask). This would be a 'femto-die' as Joseph Friedlander of Shave Shomron, Israel, has labeled it. Re-assembling these strings with perfect precision and alignment would produce more AB-matter film; leaving deliberate gaps would reproduce the 'holes' in the initial 'femto-die'.

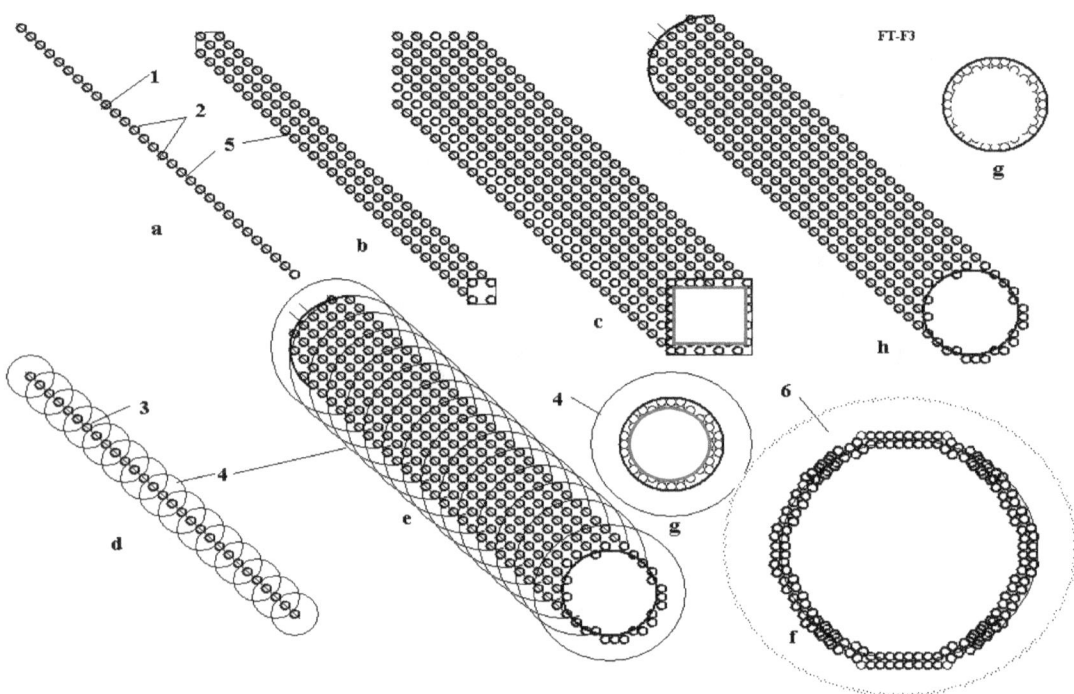

Fig. 16-2. Design of AB-Matter from nucleons (neutrons, protons, etc.) and electrons (**a**) linear one string (monofilament) (fiber, whisker, filament, thread); (**b**) ingot from four nuclear monofilaments; (**c**) multi-ingot from nuclear monofilament; (**d**) string made from protons and neutrons with electrons rotated around monofilament; (**e**) single wall femto tube (SWFT) fiber with rotated electrons; (**f**) cross-section of multi wall femto tube (MWFT) string; (**g**) cross-section of rod; (**h**) - single wall femto tube (SWFT) string with electrons inserted into AB-Matter. *Notations*: 1 – nuclear string; 2 - nucleons (neutrons, protons, etc.). 3 – protons; 4 – orbit of electrons; 5 – electrons; 6 – cloud of electrons around tube.

The developing of femtotechnology is easier, in one sense, than the developing of fully controllable nanotechnology because we have only three main particles (protons, neutrons, their ready combination of nuclei $_2$D, $_3$T, $_4$He, etc., and electrons) as construction material and developed methods of their energy control, focusing and direction.

3. Using the AB-Matter (fig.16-3). The simplest use of AB-Matter is strengthening and reinforcing conventional material by AB-Matter fiber. As it is shown in the 'Computation' section [8], AB-Matter fiber is stronger (has a gigantic ultimate tensile stress) than conventional material by a factor of millions of times, can endure millions degrees of temperature, don't accept any attacking chemical reactions. We can insert (for example, by casting around the reinforcement) AB-Matter fiber (or net) into steel, aluminum, plastic and the resultant matrix of conventional material increases in strength by thousands of times—if precautions are taken that the reinforcement stays put! Because of the extreme strength disparity design tricks must be used to assure that the fibers stay 'rooted'. The matrix form of conventional artificial fiber reinforcement is used widely in current technology. This increases the tensile stress resistance of the reinforced matrix matter by typically 2 – 4 times. Engineers dream about a nanotube reinforcement of conventional matrix materials which might increase the tensile stress by 10 – 20 times, but nanotubes are very expensive and researchers cannot decrease its cost to acceptable values yet despite years of effort.

Another way is using a construct of AB-Matter as a continuous film or net (fig.16-4).

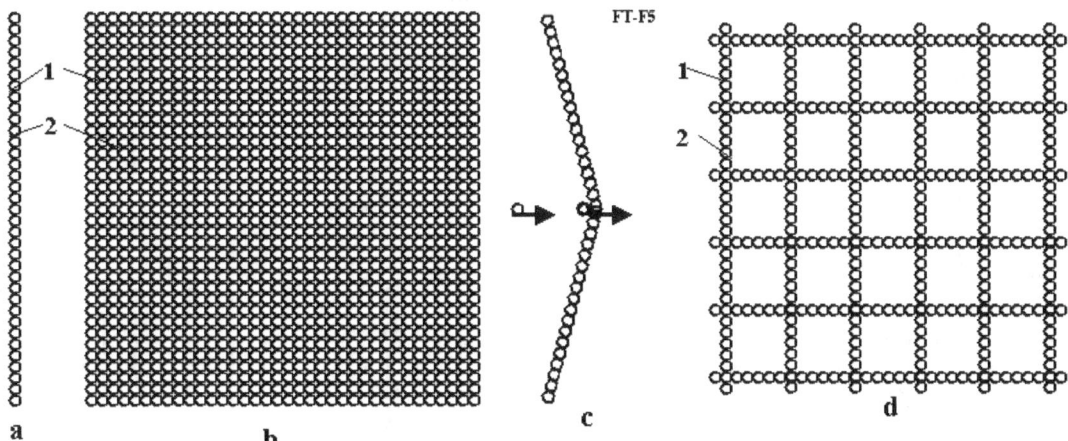

Fig. 16-3. Thin film from nuclear matter. (**a**) cross-section of a matter film from single strings (side view); (**b**) continuous film from nuclear matter; (**c**) AB film under blow from conventional molecular matter; (**d**) – net from single strings. Notations: 1 – nucleons; 2 – electrons inserted into AB-Matter; 3 – conventional atom.

These forms of AB-Matter have such miraculous properties as invisibility, superconductivity, zero friction, etc. The ultimate in camouflage, installations of a veritable Invisible World can be built from certain forms of AB-Matter with the possibility of being also interpenetable, literally allowing ghost-like passage through an apparently solid wall. Or the AB-Matter net (of different construction) can be designed as an impenetrable wall that even hugely destructive weapons cannot penetrate.

The AB-Matter film and net may be used for energy storage which can store up huge energy intensities and used also as rocket engines with gigantic impulse or weapon or absolute armor (see computation and application sections). Note that in the case of absolute armor, safeguards must be in place against buffering sudden accelerations; g-force shocks can kill even though nothing penetrates the armor!

The AB-Matter net (which can be designed to be gas-impermeable) may be used for inflatable construction of such strength and lightness as to be able to suspend the weight of a city over a vast span the width of a sea. AB-Matter may also be used for cubic or tower solid construction as it is shown in fig. 16-4.

Some Properties of AB-Matter

We spoke about the ***fantastic tensile and compressive strength, rigidity, hardness, specific strength, thermal (temperature) durability, thermal shock, and big elongation of*** AB-Matter which are more millions time then conventional matter (see [8]).

Short note about other miraculous AB-Matter properties:

1. *Zero heat/thermal capacity*. That follows because the mass of nucleons (AB-Matter string, film, net) is large in comparison with mass single atom or molecule and nucleons in AB-Matter have a very strong connection one to other. Conventional atoms and molecules cannot pass their paltry energy to AB-Matter! That would be equivalent to moving a huge dry-dock door of steel by impacting it with very light table tennis balls.

2. *Zero heat/thermal conductivity*. (See above).

3. *Absolute chemical stability. No corrosion, material fatigue. Infinity of lifetime*. All chemical reactions are

acted through ORBITAL electron of atoms. The AB-Matter does not have orbital electrons (special cases will be considered later on). Nucleons cannot combine with usual atoms having electrons. In particular, the AB-Matter has *absolute corrosion resistance*. *No fatigue of material* because in conventional material fatigue is result of splits between material crystals. No crystals in AB-Matter. That means AB-Matter has lifetime equal to the lifetime of neutrons themselves. Finally a container for the universal solvent!

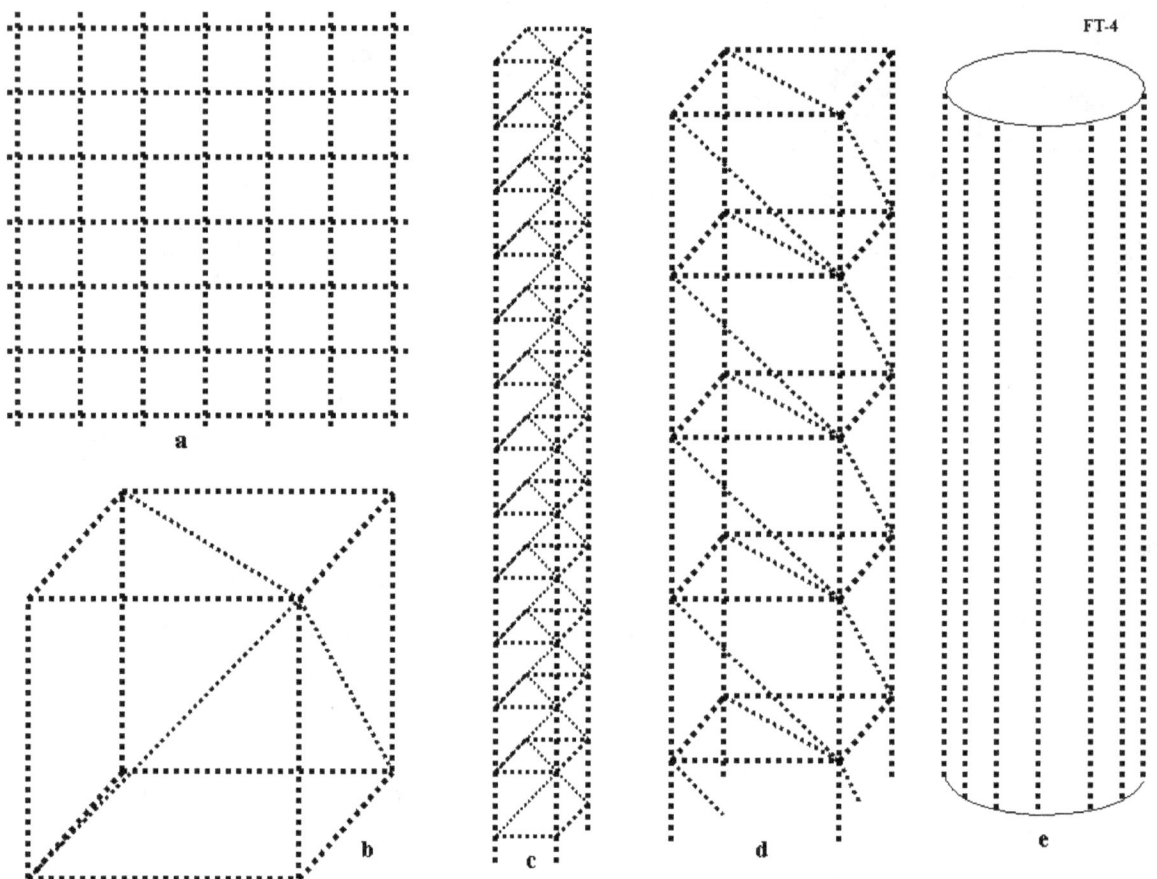

Fig.16-4. Structures from nuclear strings. (**a**) nuclear net (netting, gauze); (**b**) primary cube from matter string; (**c**) primary column from nuclear string; (**d**) large column where elements made from primary columns; (**e**) tubes from matter string or matter columns.

4. *Super-transparency, invisibility of special AB-Matter-nets.* An AB-Matter net having a step distance (mesh size) between strings or monofilaments of more than 100 fm = 10^{-13} m will pass visible light having the wave length $(400 - 800) \times 10^{-9}$ m. You can make cars, aircraft, and space ships from such a permeable (for visible light) AB-Matter net and you will see a man (who is made from conventional matter) apparently sitting on nothing, traveling with high speed in atmosphere or space without visible means of support or any visible vehicle!

5. *Impenetrability for gas, liquids, and solid bodies.* When the AB-Matter net has a step size between strings of less than atomic size of 10^{-10} m, it became impenetrabile for conventional matter. Simultaneously it may be invisible for people and have gigantic strength. The AB-Matter net may –as armor--protect from gun, cannon shells and missiles.

6. *Super-impenetrability for radiation.* If the cell size of the AB-Matter net will be less than a wave length of a given radiation, the AB-Matter net does not pass this radiation. Because this cell size may be very small, AB net is perfect protection from any radiation up to soft gamma radiation (include radiation from nuclear bomb).

7. *Full reflectivity (super-reflectivity)*. If the cell size of an AB-Matter net will be less than a wavelength of a given radiation, the AB-Matter net will then fully reflect this radiation. With perfect reflection and perfect impenetrability remarkable optical systems are possible. A Fresnel like lens might also be constructible of AB-Matter.

8. *Permeable property (ghost-like intangibility power; super-passing capacity)*. The AB-Matter net from single strings having mesh size between strings of more than 100 nm = 10^{-11} m will pass the atoms and molecules through itself because the diameter of the single string (2×10^{-15} m) is 100 thousand times less then diameter of atom (3×10^{-10} m). That means that specifically engineered constructions from AB-Matter can be built on the Earth, but people will not see and feel them. The power to phase through walls, vaults, and barriers has occasionally been portrayed in science fiction but here is a real life possibility of it happening.

9. *Zero friction*. If the AB-Matter net has a mesh size distance between strings equals or less to the atom (3×10^{-10} m), it has an ideal flat surface. That means the mechanical friction may be zero. It is very important for aircraft, sea ships and vehicles because about 90% of its energy they spend in friction. Such a perfect surface would be of vast value in optics, nanotech molecular assembly and prototyping, physics labs, etc.

10. *Super or quasi-super electric conductivity at any temperature*. As it is shown in previous section the AB-Matter string can have outer electrons in an arrangement similar to the electronic cloud into metal. But AB-Matter strings (threads) can be located along the direction of the electric intensity and they will not resist the electron flow. That means the electric resistance will be zero or very small.

11. *High dielectric strength* (see Equation (21) in [4] Ch.2).

AB-Matter may be used for devices to produce high magnetic intensity.

12. *Transfer pressure in long distance*. The pressure force of AB-string is very high and does NOT depend from its length (that will be shown in the next author research). That means we can penetrate into the human body without damage it from distance in hundreds kilometers, into the Earth (geological exploration), into other planets (Moon, Mars) without flight to them, keep the motion less satellites and build the Space Elevator from Earth surface.

Applications and new systems in Aerospace and aviation

The applications of the AB-Matter are encyclopedic in scope. This matter will create revolutions in many fields of human activity. We show only non-usual applications in aerospace, aviation that come to mind, and by no means all of these.

1. Storage of gigantic energy.

As it is shown in [3]-[7], the energy saved by flywheel equals the special mass density of material (17). As you see that is a gigantic value of stored energy because of the extreme values afforded by the strong nuclear force. Car having a pair of 1 gram counterspun fly-wheels (2 grams total) (20) charged at the factory can run all its life without benzene. Aircraft or sea ships having 100 gram (two 50 gram counterspun fly-wheels) can fly or swim all its life without additional fuel. The offered flywheel storage can has zero friction and indefinite energy storage time.

2. New propulsion system of space ship.

The most important characteristic of rocket engine is specific impulse (speed of gas or other material flow out from propulsion system). Let us compute the speed of a part of fly-wheel ejected from the offered rocket system

$$\frac{mV^2}{2} = E, \quad V = \sqrt{\frac{2E}{m}} = 3.9 \cdot 10^7 \ m/s. \quad (1)$$

Here V is speed of nucleon, m/s; E = 12.8×10^{-13} J (1) is energy of one nucleon, J; m = 1,67×10^{-27} kg is mass of one nucleon, kg. The value (1) is about 13% of light speed.

The chemical rocket engine has specific impulse about 3700 m/s. That value is 10 thousand times less. The electric rocket system has a high specific impulse but requires a powerful compact and light source of energy. In the offered rocket engine the energy is saved in the flywheel. The current projects of a nuclear rocket are very complex, heavy, and dangerous for men (gamma and neutron radiation) and have specific impulse of thousands of times less (1). The offered AB-Matter rocket engine may be very small and produced any rocket thrust in any moment in any direction.

The offered flywheel rocket engine used the AB-matter is presented in fig.16-5a. That is flywheel made from AB-matter. It has a nozzle 3 having control of exit mass. The control allows to exit of work mass in given moment and in given position of flywheel. The flywheel rotates high speed and the exhaust mass leave the rocket engine with same speed when the nozzle is open. In result the engine has thrust 6. As exhaust mass may be used any mass: liquid (for example, water), sand, small stones and other suitable planet or space material (mass). The energy needed for engine and space ship is saved in the revolving flywheel. This energy may be received at started planet or from space ship engine.

The rocket used the suggested engine is shown in fig. 5b. That has a cabin 7, the offered propulsion system 8, undercarriage 9 and rotary mechanism 10 for turning the ship in need position.

Let us to estimate the possibility of offered rocket. Notate, the relation of the exhaust mass to AM-matter cover mass of flywheel are taken a = 10, the safety (strength) factor b = 4. About 20% of space ship is payload and construction and 80% is the exhaust mass. Then exhaust speed of throw away mass and receiving speed by space ship are:

$$V = \sqrt{\frac{k}{ab}} = 2.12 \cdot 10^6 \ m/s, \quad m_s V_s = mV, \quad V_s = \frac{m}{m_s} V = \frac{0.8}{0.2} \cdot 2.12 \cdot 10^6 = 8.48 \cdot 10^6 \ m/s, \quad (2)$$

where V speed of exhausted mass, m/s; $k = \sigma/d$ = 1.9×10^{14} (m/s)^2 is strength coefficient (16); m_s is final mass of rocket, kg; V_s = 8480 km/s is final speed of rocket, m/s; m is throw off mass, kg.

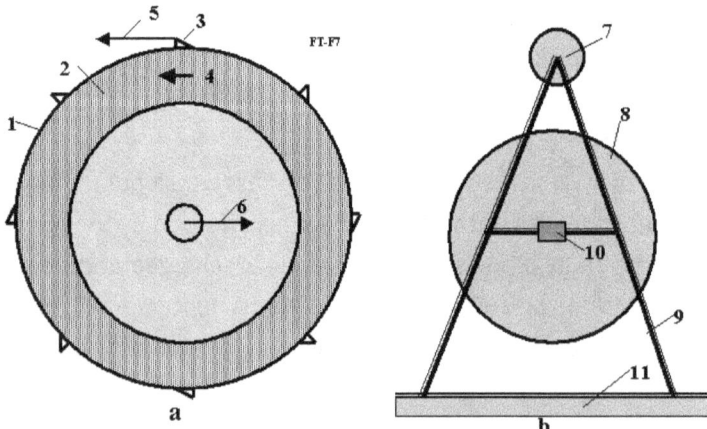

Fig. 16-5. Schema of new rocket and propulsion system. (**a**) Propulsion system from AB matter and storage energy. (**b**) Rocket with offered propulsion system.
Notations: 1 – cover (flywheel) from AB-matter; 2 – any work mass; 3 – nozzle with control of exit mass; 4 –

direction of rotation; 5 – direction of exhaust mass; 6 – thrust; 7 – space ship; 8 – offered propulsion system; 9 – undercarriage; 10 – rotary mechanism; 11 – planet surface.

3. High efficiency rocket, jet and piston aviation engines.

The efficiency conventional jet and rocket engines are very limited by the temperature and safety limits of conventional matter (2000°K). If we will design the rotor blades (in jet engine), combustion chamber (in rocket and piston engines) from AB-Matter, we radically improve their capacities and simplify their construction (for example, no necessary cooling system!).

4. Hypersonic aircraft.

The friction and heat which attacks conventional materials for hypersonic aircraft limits their speed. Using the AB-Matter deletes this problem. Many designs for aerospace planes could capture oxygen in flight, saving hauling oxidizer and carrying fuel alone—enabling airliner type geometries and payloads since the weight of the oxidizer and the tanks needed to hold it, and the airframe strengths required escalate the design and cascade through it until conventional materials today cannot build a single stage to orbit or antipodes aerospace plane. But that would be quite possible with AB-Matter.

5. Increasing efficiency of a conventional aviation and transport vehicles.

AB-Matter does not experience friction. The air drag in aviation is produced up 90% by air friction on aircraft surface. Using AB-Matter will make jump in flight characteristics of aircraft and other transport vehicles (including sea ships and cars).

6. Improving capabilities of all machines.

Appearance new high strength and high temperature AB-Matter will produce jump, technology revolution in machine and power industries.

7. Computer and computer memory.

The AB-Matter film allows to write in 1 cm^2 $N = 1/(4 \times 10^{-26}) = 2.5 \times 10^{25}$ 1/cm^2 bits information. The current 45 nanometer technology allows to write only $N = 2.5 \times 10^{14}$ 1/cm^2 bit. That means the main chip and memory of computer based in AB-Matter film may be a billion times smaller and presumably thousands of times faster (based on the lesser distance signals must travel).

The reader can imagine useful application of AB-Matter in any field he is familiar with.

8. Penetration in any body in long distance.

We can penetrate into the human body without damage it from distance in hundreds kilometers, into the interior of the Earth (geological exploration), into other planets (Moon, Mars) without flight to them, keep the motion less satellites and build the Space Elevator from Earth surface.

Discussion

1. Micro-World from AB-Matter: An Amusing Thought-Experiment. AB-Matter may have 10^{15} times more particles in a given volume than a single atom.. A human being, man made from conventional matter, contains about 5×10^{26} molecules. That means that 200 'femto-beings' of equal complexity from AB-Matter (having same number of components) could be located in the volume of one microbe having size 10 µ = 10^{-5} m. If this proved possible, we could not see them, they could not see us in terms of direct sensory input. Because of the wavelength of light it is questionable what they could learn of the observable macro-Universe. The implications, for transhuman scenarios, compact interstellar (microbe sized!) payloads, uploading and other such scenarios are profound. It is worth recalling that a single house and garden required to support a single conventional matter human is, for AB-Matter 'femto-beings', equivalent in relative vastness as the extended Solar system is for us. If such a future form could be created and minds 'uploaded' to it, the future theoretical population, knowledge base, and scholarly and knowledge-industries output of even a single planet so populated could rival that of a theoretical Kardashev Type III galactic civilization!

2. Stability of AB-matter.

Readers usually ask: what is the connection (proton to proton) given a new element when, after 92 protons, the connection is unstable?

Answer: That depends entirely on the type of connection. If we conventionally join the carbon atom to another carbon atom a lot of times, we then get the conventional piece of a coil. If we joint the carbon atom to another carbon atom by the indicated special methods, we then get the very strong single-wall nanotubes, graphene nano-ribbon (super-thin film), armchair, zigzag, chiral, fullerite, torus, nanobud and other forms of nano-materials. That outcome becomes possible because the atomic force (van der Waals force, named for the Dutch physicist Johannes Diderik van der Waals, 1837-1923, etc.) is NON-SPHERICAL and active in the short (one molecule) distance. The nucleon nuclear force also is NON-SPHERICAL and they may also be active about the one nucleon diameter distance (Fig. 16-1). That means we may also produce with them the strings, tubes, films, nets and other geometrical constructions.

The further studies (it will be published) are shown that AB-matter will be stability if:
1) The any sphere having radius $R \approx 6 \times 10^{-15}$ m in any point of structure figs. 1 – 4 must contain NOT more 238 nucleons (about 92 of them must be protons). That means any cross-section area of the solid rod, beam and so on of AB-structure (for example figs. 16-1b,c,g) must contain NOT more about 36 nucleons.
2) AB-matter must contains the proton in a certain order because the electrostatic repel forces of them give the stability of the given structure.

Conclusion

The author offers a design for a new form of nuclear matter from nucleons (neutrons, protons), electrons, and other nuclear particles. He shows that the new AB-Matter has most extraordinary properties (for example, (in varying circumstances) remarkable tensile strength, stiffness, hardness, critical temperature, superconductivity, super-transparency, ghostlike ability to pass through matter, zero friction, etc.), which are millions of times better than corresponded properties of conventional molecular matter. He shows how to design aircraft, ships, transportation, thermonuclear reactors, and constructions, and so on from this new nuclear matter. These vehicles will have correspondingly amazing possibilities (invisibility, passing through any walls and amour, protection from nuclear bombs and any radiation, etc).

People may think this fantasy. But fifteen years ago most people and many scientists thought – nanotechnology is fantasy. Now many groups and industrial labs, even startups, spend hundreds of millions of dollars for development of nanotechnological-range products (precise chemistry, patterned atoms, catalysts, metamaterials, etc) and we have nanotubes (a new material which does not exist in Nature!) and other achievements beginning to come out of the pipeline in prospect. Nanotubes are stronger than steel by a hundred times—surely an amazement to a 19th Century observer if he could behold them.

Nanotechnology, in near term prospect, operates with objects (molecules and atoms) having the size in nanometer (10^{-9} m). The author here outlines perhaps more distant operations with objects (nuclei) having size in the femtometer range, (10^{-15} m, millions of times less smaller than the nanometer scale). The name of this new technology is femtotechnology.

I want to explain the main thrust of this by analogy. Assume we live some thousands of years ago in a great river valley where there are no stones for building and only poor timber. In nature we notice that there are many types of clay (nuclei of atom—types of elemet). One man offers to people to make from clay bricks (AB-Matter) and build from these bricks a fantastic array of desirable structures too complex to make from naturally occuring mounds of mud. The bricks enable by increased precision and strength things impossible before. A new level of human civilization begins.

I call upon scientists and the technical community to to research and develop femtotechnology. I think we can reach in this field progress more quickly than in the further prospects of nanotechnology, because we have fewer (only 3) initial components (proton, neutron, electron) and interaction between them is well-known (3 main forces: strong, weak, electostatic). The different conventional atoms number about 100, most commone moleculs are tens thousands and interactions between them are very complex (e.g. Van der Waals force).

It may be however, that nano and femto technology enable each other as well, as tiny bits of AB-Matter would be marvellous tools for nanomechanical systems to wield to obtain effects unimaginable otherwise.

What time horizon might we face in this quest? The physicist Richard Feynman offeredhis idea to design artificial matter from atoms and molecules at an American Physical Society meeting at Caltech on December 29, 1959. But only in the last 15 years we have initial progress in nanotechnology. On the other hand progress is becoming swifter as more and better tools become common and as the technical community grows.

Now are in the position of trying to progress from the ancient 'telega' haywagon of rural Russia (in analogy, conventional matter composites) to a 'luxury sport coupe' (advanced tailored nanomaterials). The author suggests we have little to lose and literal worlds to gain by simultaneously researching how to leap from 'telega' to 'hypersonic space plane'. (Femotech materials and technologies, enabling all the wonders outlined here).

References

Many works noted below the reader can find on site Cornel University <http://arxiv.org/>, sites <http://vixra.org> , <http://www.archive.org> search "Bolonkin", <http://bolonkin.narod.ru/p65.htm> and in Conferences 2002-2006 (see, for example, Conferences AIAA, <http://aiaa.org/> , search "Bolonkin").

1. Bolonkin A.A., (2006)"Non-Rocket Space Launch and Flight", Elsevier, 2005, http://www.archive.org/details/Non-rocketSpaceLaunchAndFlight,

2. Bolonkin A.A. (2006), Book "*New Concepts, Ideas and Innovation in Aerospace*", NOVA, 2008.
http://www.archive.org/details/NewConceptsIfeasAndInnovationsInAerospaceTechnologyAndHumanSciences

3. Bolonkin A.A. (2007), "Macro-Engineering: Environment and Technology", pp. 299-334, NOVA, 2008. http://Bolonkin.narod.ru/p65.htm,
http://www.archive.org/details/Macro-projectsEnvironmentsAndTechnologies

4. Bolonkin A.A. (2008), "New Technologies and Revolutionary Projects", Scribd, 2008, 324 pgs, http://www.archive.org/details/NewTechnologiesAndRevolutionaryProjects

5. Bolonkin A.A., (2006). Electrostatic AB-Ramjet Space Propulsion. http://arxiv.org/ftp/physics/papers/0701/0701073.pdf

6. Bolonkin A.A., (2009). Magnetic Space AB-Accelerator. http://www.scribd.com/doc/26885058

7. Bolonkin A.A., (2009). Converting of any Matter to Nuclear Energy by AB-Generator and Aerospace . **http://www.archive.org/details/ConvertingOfAnyMatterToNuclearEnergyByAb-generatorAndAerospace**, American Journal of Engineering and Applied Science, Vol. 2, #4, 2009, pp.683-693. http://viXra.org/abs/1309.0200, https://www.scribd.com/search-documents?query=Bolonkin, https://www.academia.edu/14515398/Converting_of_Matter_to_Nuclear_Energy

8. Bolonkin A.A., (2009).) **Femtotechnology: Design of the Strongest AB-Matter for Aerospace.** Presented as paper AIAA-2009-4620 to 4 Joint Propulsion Conference, 2-5 August, 2009, Denver CO, USA. See also closed paper AIAA-2010-1556 in 48 Aerospace Meeting, New Horizons, 4 – 7 January, 2010, Orlando, FL, USA. Global Science Journal, http://gsjournal.net/Science-Journals/Research%20Papers-Engineering%20%28Applied%29/Download/5713 ,
Journal of Aerospace Engineering, Oct. 2010, Vol. 23, No. 4, pp.281-292. http://viXra.org/abs/1401.0173 , http://www.archive.org/details/FemtotechnologyDesignOfTheStrongestAb-matterForAerospace, http://intellectualarchive.com #1362 . http://gsjournal.net/Science-Journals/Essays/View/5713 , https://www.academia.edu/14515249/FEMTOTECHNOLOGY_THE_STRONGEST_AB-MATTER_FOR_AEROSPACE

9. Bolonkin A.A., Book "Non-Rocket Space Launch and Flight", Elsevier. 2006, Ch. 9 "*Kinetic Anti-Gravotator*",pp. 165-186; http://Bolonkin.narod.ru/p65.htm ; Main idea of this Chapter was presented as papers COSPAR-02, C1.1-0035-02 and IAC-02-IAA.1.3.03, 53rd International Astronautical Congress. The World Space Congress-2002, 10-19 October 2002, Houston, TX, USA, and the full manuscript was accepted as AIAA-2005-4504, 41st Propulsion Conference, 10-12 July 2005, Tucson, AZ, USA.http://aiaa.org search "Bolonkin".

10. Bolonkin A.A., Book "Non-Rocket Space Launch and Flight", Elsevier. 2006, Ch.5 "*Kinetic Space Towers*", pp. 107-124, Springer, 2006. http://Bolonkin.narod.ru/p65.htm or http://www.archive.org/details/Non-rocketSpaceLaunchAndFlight

11. Bolonkin A.A., "Transport System for Delivery Tourists at Altitude 140 km", manuscript was presented as paper IAC-02-IAA.1.3.03 at the World Space Congress-2002, 10-19 October, Houston, TX, USA. http://Bolonkin.narod.ru/p65.htm ,

12. Bolonkin A.A. (2003), "Centrifugal Keeper for Space Station and Satellites", JBIS, Vol.56, No. 9/10, 2003, pp. 314-327. http://Bolonkin.narod.ru/p65.htm . See also [10] Ch.3.

13. Bolonkin A.A., Book "Non-Rocket Space Launch and Flight", by A.Bolonkin, Elsevier. 2006, Ch.3 "*Circle Launcher and Space Keeper*", pp.59-82 , http://www.archive.org/details/Non-rocketSpaceLaunchAndFlight

14. Bolonkin A.A., Book "New Concepts, Ideas and Innovation in Aerospace", NOVA, 2008, Ch. 3 " *Electrostatic AB-Ramjet Space Propulsion)*", pp.33-66.
http://www.archive.org/details/NewConceptsIfeasAndInnovationsInAerospaceTechnologyAndHumanSciences

15. Bolonkin A.A., Book "New Concepts, Ideas and Innovation in Aerospace", NOVA, 2008, Ch.12, pp.205-220. Ch. "*AB Levitrons and Their Applications to Earth's Motionless Satellites*".
http://www.archive.org/details/NewConceptsIfeasAndInnovationsInAerospaceTechnologyAndHumanSciences

16. Bolonkin A.A., Book "Macro-Projects: Environment and Technology", NOVA, 2008, Ch.10, pp.199-226," *AB-Space Propulsion*", http://Bolonkin.narod.ru/p65.htm .
http://www.archive.org/details/Macro-projectsEnvironmentsAndTechnologies .

17. Bolonkin A.A., Magnetic Suspended AB-Structures and Moveless Space Satellites.
http://www.scribd.com/doc/25883886 , Book: Bolonkin A.A., Femtotechnologies and Innovative Projects, USA, Lulu. http://viXra.org/abs/1309.0191 .

18. BolonkinA.A., Krinker M., Magnetic-Space-Launcher.
http://arxiv.org/ftp/arxiv/papers/0903/0903.5008.pdf

21. Bolonkin A.A., LIFE. SCIENCE. FUTURE (Biography notes, researches and innovations). Scribd, 2010, 208 pgs. 16 Mb. http://www.scribd.com/doc/48229884,
http://www.archive.org/details/Life.Science.Future.biographyNotesResearchesAndInnovations

Hypersonic aircraft

Part C

Popular Reviews of New Consepts, Ideas, Innovation in Space Launch and Flight

Abstract

In the past years the author and other scientists have published a series of new methods which promise to revolutionize the space technology. These include the Space Elevator, Men without the space suite into space, Artificial gravity, New method of atmospheric re-entry for space ship, Inflatable Dome for Moon, Mars, asteroids, Closed loop water cycle, Climber for Space Elevator, Cheap Protection from Nuclear Warhead, Wireless transfer of electricity throw outer Space, Artificial explosion of Sun, etc.

Some of them have the potential to decrease the space research costs in thousands of time, other allow decreasing the cost of the space exploration.

The author reviews and summarizes some revolutionary ideas, innovations and patent applications for scientists, engineers, inventors, students and the public.

Key words: Space Elevator, Men without the space suite into space, Artificial gravity, New method of atmospheric re-entry, Inflatable Dome for space, Closed loop water cycle, Climber for Space Elevator, Cheap protection from Nuclear Warhead, Wireless transfer of electricity throw outer Space, Artificial explosion of Sun.

Introduction

Space technology is technology that is related to entering, and retrieving objects or life forms from space.

"Every day" technologies such as weather forecasting, remote sensing, GPS systems, satellite television, and some long distance communications systems critically rely on space infrastructure. Of sciences astronomy and Earth sciences (via remote sensing) most notably benefit from space technology.

Computers and telemetry were once leading edge technologies that might have been considered "space technology" because of their criticality to boosters and spacecraft. They existed prior to the Space Race of the Cold War (between the USSR and the USA.) but their development was vastly accelerated to meet the needs of the two major superpowers' space programs. While still used today in spacecraft and missiles, the more prosaic applications such as remote monitoring (via telemetry) of patients, water plants, highway conditions, etc. and the widespread use of computers far surpasses their space applications in quantity and variety of application.

Space is such an alien environment that attempting to work in it requires new techniques and knowledge. New technologies originating with or accelerated by space-related endeavors are often subsequently exploited in other economic activities. This has been widely pointed to as beneficial by space advocates and enthusiasts favoring the investment of public funds in space activities and programs. Political opponent[)] counter that it would be far cheaper to develop specific technologies directly if they are beneficial and scoff at this justification for public expenditures on space-related research.

After World War 2 the space technology have received the great progress and achieved a great success. But space technology are very expensive and have limited possibilities. In the beginning 21th century the researches of some revolutionary space technology started [1]-[22]. These new technology which promise to decrease the cost of a space exploration in hundreds times. Some of them are described in this review.

Current status of new space technology and systems. Over recent years interference-fit joining technology including the application of space methods has become important in the achievement of space propulsion system. Part results in the area of non-rocket space launch and flight methods have been patented recently or are patenting now.

Professor Bolonkin made a significant contribution to the study of the new revolutionary space technology in recent years [1]-[22] (1982-2011). Some of them are presented in given review.

Space Elevator, Transport System for Space Elevator is researched in [1] Ch.1; Men without the space suite into space is described in [1] Ch.19; Electrostatic levitation and Artificial gravity is investigated in [1] Ch.15; New method of atmospheric re-entry of space ship was studied in [2] Ch.8; Inflatable Dome for Moon, Mars, satellites, and space hotel is described [3] Ch2; AB method irrigation for planet without water (Closed loop water cycle) is studied in [3] Ch.1; Artificial explosion of the Sun was researched in [4] Ch.10.

Some of these systems were developed in [5]-[23].

Many useful ideas and innovations for space technologies in given field were presented in patent and patent application section of References. In particularly, there are:

Significant scientific, interplanetary and industrial use did not occur until the 20th century, when rocketry was the enabling technology of the Space Age, including setting foot on the Moon.

But rockets are very expensive and have limited possibilities. In the beginning 21th century the researches of new space technologies started []-[22].Some of them are described in this review.

Main types of Non-Rocket Space Propulsion System

1. Space Elevator, Transport System for Space Elevator
2. Men without the space suite into space,
3. Electrostatic levitation and Artificial gravity,
4. New method of atmospheric re-entry of space ship,
5. Inflatable Dome for Moon, Mars, satellites, and space hotel
6. Ab method irrigation for planet without water. (Closed loop water cycle), |
7. Artificial explosion of the Sun.

1. Space Elevator, Transport System for Space Elevator*

The reseach brings together research on the space elevator and a new transportation system for it. This transportation system uses mechanical energy transfer and requires only minimal energy so that it provides a "Free Trip" into space. It uses the rotary energy of planets. The research contains the theory and results of computations for the following projects: 1. Transport System for Space Elevator. The low cost project will accommodate 100,000 tourists annually. 2. Delivery System for Free Round Trip to Mars (for 2000 people annually). 3 Free Trips to the Moon (for 10,000 tourists annually).

The projects use artificial material like nanotubes and whiskers that have a ratio of strength to density equal to 4 million meters. At present scientific laboratories receive nanotubes that have this ratio equal to 20 million meters.

----

* That part of the chapter was presented by author as paper IAC-02-V.P.07 at the World Space Congress-2002, Oct.10-19, Houston, TX, USA and published in *JBIS*, vol. 56, No. 7/8, 2003, pp. 231–249. See also Bolonkin A.A., (2005)"Non-Rocket Space Launch and Flight", Elsevier, 2005, Ch.1, http://www.archive.org/details/Non-rocketSpaceLaunchAndFlight,

Free trip to Space (Project 1)

Description

A proposed centrifugal space launcher with a cable transport system is shown in Fig.1-1. The system includes an equalizer (balance mass) located in geosynchronous orbit, an engine located on Earth, and the cable transport system having three cables: a main (central) cable of equal stress, and two transport cables, which include a set of mobile cable chains connected sequentially one to an other by the rollers. One end of this set is connected to the equalizer, the other end is connected to the planet. Such a separation is necessary to decrease the weight of the transport cables, since the stress is variable along the cable. This transport system design requires a minimum weight because at every local distance the required amount of cable is only that of the diameter for the local force. The load containers are also connected to the chain. When containers come up to the rollers, they move past the rollers and continue their motion up the cable. The entire transport system is driven by any conventional motor located on the planet. When payloads are not being delivered into space, the system may be used to transfer mechanical energy to the equalizer (load cabin, the space station). This mechanical energy may also be converted to any other sort energy.

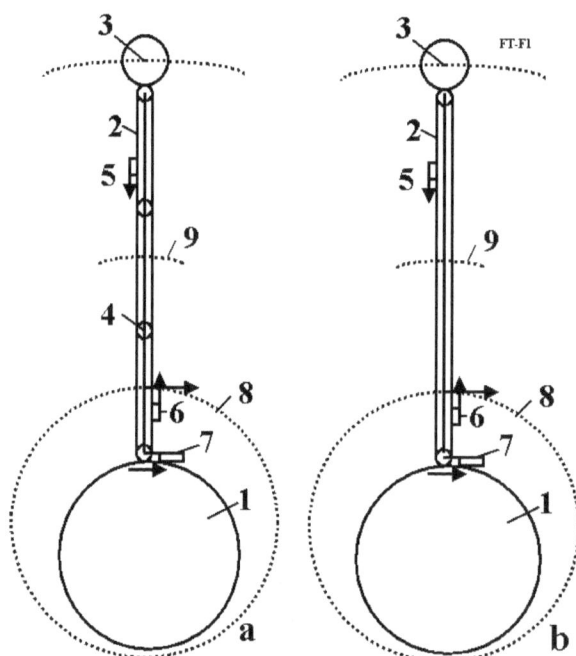

Fig. 1-1a,b. The suggested Space Transport System. Notations: 1 – Rotary planet (for example, the Earth); 2 - suggested Space Transport System ; 3 - equalizer (counterweight); 4 - roller of Transport System; 5 - launch space ship; 6 - a return ship after flight ; 7 – engine of Transport System; 8 – elliptic orbit of tourist vehicles; 9 - Geosynchronous orbit. a – System for low coefficient k, b – System for high coefficient k (without rollers 4).

The space satellites released below geosynchronous orbit will have elliptic orbits and may be connected back to the transport system after some revolutions when the space ship and cable are in the same position (Fig.1- 1). If low earth orbit satellites use a brake parachute, they can have their orbit closed to a circle.

The space probes released higher than geosynchronous orbit will have a hyperbolic orbit, fly to other planets, and then can connect back to the transport system when the ship returns.

Most space payloads, like tourists, must be returned to Earth. When one container is moved up, then another container is moved down. The work of lifting equals the work of descent, except for a small loss in the upper and lower rollers. The suggested transport system lets us fly into space without expending enormous energy. This is the reason why the method and system are named a "Free Trip".

Devices shown on fig. 1-2 are used to change the cable length (or chain length). The middle roller is shown in fig. 1-3.

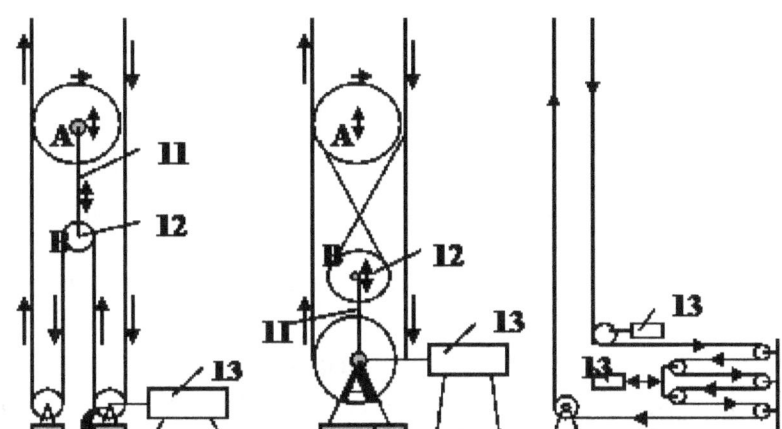

Fig. 1-2. Two mechanisms for changing the rope length in the Transport System (They are same for the space station). Notations: 11 - the rope which is connected axis A,B. This rope can change its length (the distance AB); 12 - additional rollers.

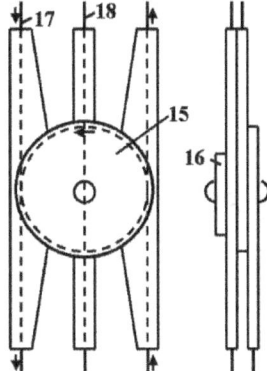

Fig. 1-3. Roller of Space Transport System. Notations: 15 – roller, 16 – control; 17 – transport system cable; 18 – main cable.

If the cable material has a very high ratio of safe (admissible) stress/density there may be one chain (Fig. 1-1b). The transport system then has only one main cable. Old design (fig. 1-1.1) has many problems, for example, in the transfer of large amounts of energy to the load cabin.

Theory and Computation of optimal cable

The computation for different $K=\sigma/\gamma/10^7$ is presented in [1] p.11, Fig. 1.6. Here σ is tensile strength of cable

[N/m²], γ_o is density of cable [kg/m³].

Transport system for Space Elevator ([1] Project 1)

That is an example of an inexpensive transport system for cheap annual delivery of 100,000 tourists, or 12,000 tons of payload into Earth orbits, or the delivery up to 2,000 tourists to Mars, or the launching of up to 2,500 tons of payload to other planets.

Main results of computation

The suggested space transport system can be used for delivery of tourists and payloads to an orbit around the Earth, or to space stations serving as a tourist hotel, scientific laboratory, or industrial factory, or for the delivery of people and payloads to other planets.

Technical parameters: Let us take the safe cable stress 7200 kg/mm² and cable density 1800 kg/m³. This is equal to $K = 4$. This is not so great since by the year 2000 many laboratories had made experimental nanotubes with a tensile stress of 200 Giga–Pascals (20,000 kg/mm²) and a density of 1800 kg/m³. The theory of nanotubes predicts 100 ton/mm² with Young's modulus of up to 5 Tera Pascal's (currently it is 1 Tera Pascal) and a density of 800 kg/m³ for SWNTs nanotubes. This means that the coefficient K used in our equations and graphs can be up to 125.

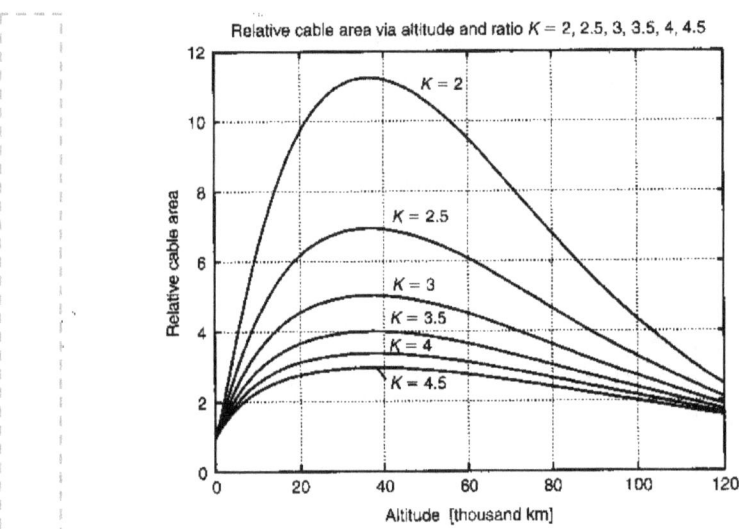

Fig. 1.6. Relative cable area via altitude [thousand km] for coefficient $K = 2\text{–}4.5$.

[1] Fig. 1.6.

Assume a maximum equalizer lift force of 9 tons at the Earth's surface and divide this force between three cables: one main and two transport cables. Then it follows from [1] Fig. 1.11, that the mass of the equalizer (or the space station) creates a lift force of 9 tons at the Earth's surface, which equals 518 tons for $K = 4$ (this is close to the current International Space Station weight of 450 tons). The equalizer is located over a geosynchronous orbit at an altitude of 100,000 km. Full centrifugal lift force of the equalizer ([1] Fig. 1.10) is 34.6 tons, but 24.6 tons of the equalizer are used in support of the cables. The transport system has three cables: one main and two in the transport system. Each cable can support a force (load) of 3000 kgf. The main cable has a cross-sectional area of equal stress. Then the cable cross-section area is (see [1] Fig. 1.6) $A = 0.42$ mm² (diameter $D = 0.73$ mm) at the Earth's surface, maximum 1.4 mm² in the middle section ($D = 1.33$ mm, altitude 37,000 km), and $A = 0.82$ mm² ($D = 1$ mm) at the equalizer. The mass of main cable is 205 tons (see [1] Fig.

1.8). The chains of the two transport cable loops have gross section areas to equal the tensile stress of the main cable at given altitude, and the capabilities are the same as the main cable. Each of them can carry 3 tons force. The total mass of the cable is about 620 tons. The three cables increase the safety of the passengers. If any one of the cables breaks down, then the other two will allow a safe return of the space vehicle to Earth and the repair of the transport system.

If the container cable is broken, the pilot uses the main cable for delivering people back to Earth. If the main cable is broken, then the load container cable will be used for delivering a new main cable to the equalizer. For lifting non-balance loads (for example, satellites or parts of new space stations, transport installations, interplanetary ships), the energy must be spent in any delivery method. This energy can be calculated from equation [1] (1.8) (Fig. 1.15). When the transport system in Fig. 1-1 is used, the engine is located on the Earth and does not have an energy limitation [1][11]. Moreover, the transport system in Fig. 1-1 can transfer a power of up to 90,000 kW to the space station for a cable speed of 3 km/s. At the present time, the International Space Station has only 60 kW of power.

Delivery capabilities. For tourist transportation the suggested system works in the following manner. The passenger space vehicle has the full mass of 3 tons (6667 pounds) to carry 25 passengers and two pilots. One ship moves up, the other ship, which is returning, moves down; then the lift and descent energies are approximately equal. If the average speed is 3 km/s then the first ship reaches the altitude of 21.5 – 23 thousands km in 2 hours (acceleration 1.9 m/s^2). At this altitude the ship is separated from the cable to fly in an elliptical orbit with minimum altitude 200 km and period approximately 6 hours ([1] Figs. 1.16, 1.17). After one day the ship makes four revolutions around the Earth while the cable system makes one revolution, and the ship and cable will be in the same place with the same speed. The ship is connected back to the transport system, moves down the cable and lifts the next ship. The orbit may be also 3 revolutions (period 8 hours) or 2 revolutions (period 12 hours). In one day the transport system can accommodate 12 space ships (300 tourists) in both directions. This means more then 100,000 tourists annually into space.

The system can launch payloads into space, and if the altitude of disconnection is changed then the orbit is changed (see [1] Fig. 1.17). If a satellite needs a low orbit, then it can use the brike parachute when it flies through the top of the atmosphere and it will achieve a near circular orbit. The annual payload capability of the suggested space transport system is about 12,600 tons into a geosynchronous orbit.

If instead of the equalizer the system has a space station of the same mass at an altitude of 100,000 km and the system can has space stations along cable and above geosynchronous orbit then these stations decrease the mass of the equalizer and may serve as tourist hotels, scientific laboratories, or industrial factories.

If the space station is located at an altitude of 100,000 km, then the time of delivery will be 9.36 hours for an average delivery speed of 3 km/s. This means 60 passengers per day or 21,000 people annually in space.

Let us assume that every person needs 400 kg of food for a one–year round trip to Mars, and Mars has the same transport installation (see next project). This means we can send about 2000 people to Mars annually at suitable positions of Earth relative to Mars.

Estimations of installation cost and production cost of delivery

Cost of suggested space transport installation [1][5,6]. The current International Space Station has cost many billions of dollars, but the suggested space transport system can cost a lot less. Moreover, the suggested transport system allows us to create other transport systems in a geometric progression [see equation [1](1.13)]. Let us examine an example of the transport system.

Initially we create the transport system to lift only 50 kg of load mass to an altitude of 100,000 km. Using the [1] Figs. 1.6 to 1.14 we have found that the equalizer mass is 8.5 tons, the cable mass is 10.25 tons and the total mass is about 19 tons. Let us assume that the delivery cost of 1 kg mass is $10,000. The construction of the system will then have a cost of $190 million. Let us assume that 1 ton of cable with $K = 4$ from whiskers or nanotubes costs $0.1 million then the system costs $1.25 million. Let us put the research and development (R&D) cost of installation at $29 million. Then the total cost of initial installation will be $220 million. About 90% of this sum is the cost of initial rocket delivery.

After construction, this initial installation begins to deliver the cable and equalizer or parts of the space station into space. The cable and equalizer capability increase in a geometric progression. The installation can use part of the time for delivery of payload (satellites) and self-financing of this project. After 765 working days the total mass of equalizer and cables reaches the amount above (1133 tons) and the installation can work full time as a tourist launcher or continue to create new installations. In the last case this installation and its derivative installations can build 100 additional installations (1133 tons) in only 30 months [see equation [1] (1.13) and Fig. 1.21] with a total capability of 10 million tourists per year. The new installations will be separated from the mother installations and moved to other positions around the Earth. The result of these installations allows the delivery of passengers and payloads from one continent to another across space with low expenditure of energy.

Let us estimate the cost of the initial installation. The installation needs 620 tons of cable. Let us take the cost of cable as $0.1 million per ton. The cable cost will be $62 million. Assume the space station cost $20 million. The construction time is 140 days [equation (1.13)]. The cost of using of the mother installation without profit is $5 millions/year. In this case the new installation will cost $87 million. In reality the new installation can soon after construction begin to launch payloads and become self-financing.

Cost of delivery

The cost of delivery is the most important parameter in the space industry. Let us estimate it for the full initial installation above.

As we calculated earlier the cost of the initial installation is $220 millions (further construction is made by self-financing). Assume that installation is used for 20 years, served by 100 officers with an average annual salary of $50,000 and maintenance is $1 million in year. If we deliver 100,000 tourists annually, the production delivery cost will be $160/person or $1.27/kg of payload. Some 70% of this sum is the cost of installation, but the delivery cost of the new installations will be cheaper.

If the price of a space trip is $1990, then the profit will be $183 million annually. If the payload delivery price is $15/kg then the profit will $189 millions annually.

The cable speed for $K = 4$ is 6.32 km/s [equation [1] (1.11), Fig. 1.19]. If average cable speed equals 6 km/s, then all performance factors are improved by a factor of two times.

If the reader does not agree with this estimation, then equations [1] (1.1) to (1.13) and Figs. 1.6 to 1.21 are able calculation of the delivery cost for other parameters. In any case the delivery cost will be hundreds of times less than the current rocket powered method.

Delivery System for Free Round Trip to Mars (Project 2)

A method and similar installation Fig.1-1, [1](Figs.1 to 4) can be used for inexpensive travel to other planets, for example, from the Earth to Mars or the Moon and back [1](Fig. 1.22). A Mars space station would be similar to an Earth space station, but the Mars station would weigh less due to the decreased gravitation on Mars. This method uses the rotary energy of the planets. For this method, two facilities are required, one on Earth and the other on another planet (e.g. Mars). The Earth accelerates the space ship to the required speed and direction and then disconnects the ship. The space ship flies in space along the defined trajectory to Mars (Fig.1-4). On reaching Mars the space ship connects to the cable of the Mars space transport system, then it moves down to Mars using the transport system.

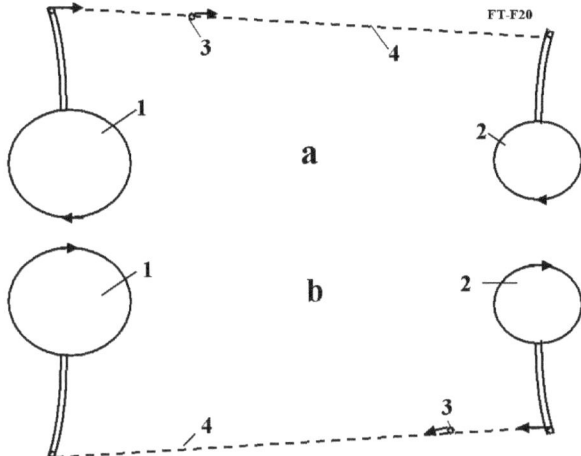

Fig. 1-4. Using the suggested transport system for space flight to Mars and back. Notation:
1 – Earth, 2 – Mars, 3 – space ship, 4 – trajectory of space ship to Mars (a) and back (b).

Free Trip to Moon (Project 3)

This method may be used for an inexpensive trip to a planet's moon, if the moon's angular speed is equal to the planet's angular speed, for example, from the Earth to the Moon and back (Fig. 1-5 to 1-7). The upper end of the cable is connected to the planet's moon. The lower end of the cable is connected to an aircraft (or buoy), which flies (i.e. glides or slides) along the planet's surface. The lower end may be also connected to an Earth pole. The aircraft (or Earth polar station, or Moon) has a device which allows the length of cable to be changed. This device would consist of a spool, motor, brake, transmission, and controller. The facility could have devices for delivering people and payloads to the Moon and back using the suggested transport system. The delivery devices include: containers, cables, motors, brakes, and controllers. If the aircraft is small and the cable is strong then the motion of the Moon can be used to move the airplane. For example, if the airplane weighs 15 tons and has an aerodynamic ratio (the lift force to the drag force) equal to 5, a thrust of 3000 kg would be enough for the aircraft to fly for infinity without requiring any fuel. The aircraft could use a small engine for maneuverability and temporary landing. If the planet has an atmosphere (as the Earth) the engine could be a turbine engine. If the planet does not have an atmosphere, a rocket engine may be used.

If the suggested transport system is used only for free thrust (9 tons), the system can thrust the three named supersonic aircraft or produce up to 40 millions watts of energy.

A different facility could use a transitional space station located at the zero gravity point between the planet and the planet's moon. Fig. 6 shows a sketch of the temporary landing of an airplane on the planet surface. The aircraft increases the length of the cable, flies ahead of the cable, and lands on a planet surface. While the planet makes an angle turn ($\alpha + \beta = 30º$, see Fig. 1.31) the aircraft can be on a planet surface. This time equals about 2 hours for the Earth, which would be long enough to load payload on the aircraft.

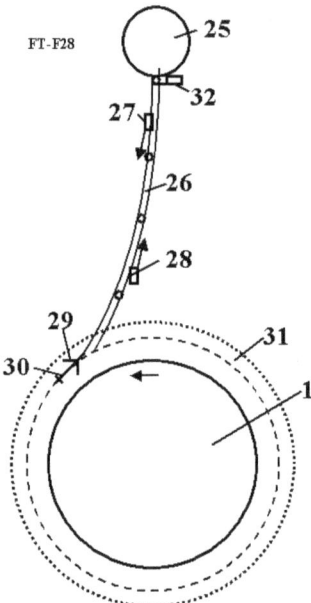

Fig.1- 5. The suggested transport system for the Moon. Notations: 1 – Earth, 25 - Moon, 26 – suggested Moon transport system, 27, 28 – load cabins, 29 – aircraft, 30 – cable control, 32 – engine.

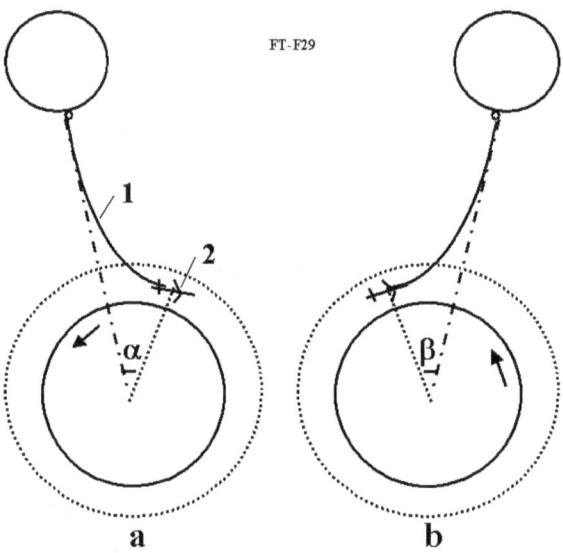

Fig.1- 6. Temporary landing of the Moon aircraft on the Earth's surface for loading. a– landing, b– take-off.

The Moon's trajectory has an eccentricity (Fig.1-7). If the main cable is strong enough, the moon may used to pull a payload (space ship, manned cabin), by trajectory to an altitude of about 60,000 kilometers every 27 days. For this case, the length of the main cable from the Moon to the container does not change and when the Moon increases its distance from the Earth, the Moon lifts the space ship. The payload could land back on the planet at any time if it is allowed to slide along the cable. The Moon's energy can be used also for an inexpensive trip around the Earth (Figs. 1-5 and 1-7) by having the moon "drag" an aircraft around the planet (using the Moon as a free thrust engine). The Moon tows the aircraft by the cable at supersonic speed, about 440 m/s (Mach number is 1.5).

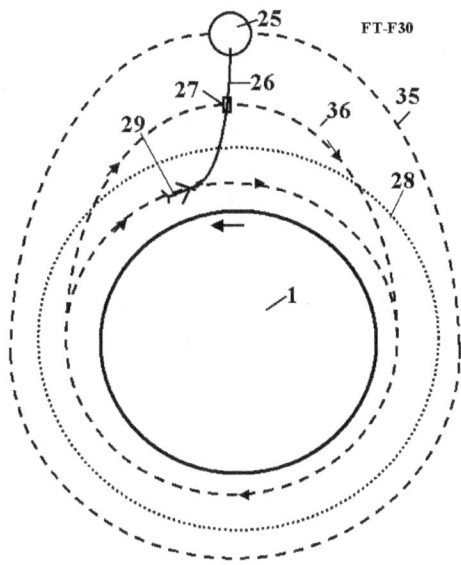

Fig. 1-7. Using the Moon's elliptical orbit for a free trip in space of up to 71,000 km. Notations: 1 – Earth, 25 – Moon, 26 – cable from Earth to Moon, 27 – Space Vehicle, 28 – limit of Earth atmosphere, 35 – Moon orbit, 36 – elliptical orbit of a Moon vehicle.

The other more simple design (without aircraft) is shown in [1] Fig. 7.1, chapter 7. The cable is connected to on Earth pole, to a special polar station which allows to change a length of cable. Near the pole the cable is supported in the atmosphere by air balloons and wings.

Technical parameters

The following are some data for estimating the main transport system parameters for connecting to the Moon to provide inexpensive payload transfer between the Earth and the Moon. The system has three cables, each of which can keep the force at 3 tons. Material of the cable has $K=4$. All cables would have cross-sectional areas of equal stress. The cable has a minimal cross-sectional area A_0 of 0.42 mm² (diameter d = 0.73 mm) and maximum cross-sectional area A_m of 1.9 mm² (d = 1.56 mm). The mass of the main cable would be 1300 tons [1] (Fig. 1.36). The total mass of the main cable plus the two container cables (for delivering a mass of 3000 kg) equals 3900 tons for the delivery transport system in [1] Figs. 1.30 to 1.33. An inexpensive means of payload delivery between the Earth and the Moon could thus be developed. The elapsed time for the Moon trip at a speed of 6 km/s would be about 18.5 hours and the annual delivery capability would be 1320 tons in both directions.

Discussion
Cable Problems

Most engineers and scientists think it is impossible to develop an inexpensive means to orbit to another planet. Twenty years ago, the mass of the required cable would not allow this proposal to be possible for an additional speed of more 2,000 m/s from one asteroid. However, today's industry widely produces artificial fibers that have a tensile strength 3–5 times more than steel and a density 4–5 times less than steel. There are also experimental fibers which have a tensile strength 30–60 times more than steel and a density 2 to 4 times less than steel. For example, in the book *Advanced Fibers and Composites* is p. 158, there is a fiber C_D with a tensile strength of σ = 8000 kg/mm² and density (specific gravity) γ = 3.5 g/cm³. If we take an admitted strength of 7000 kg/mm² ($\sigma = 7 \times 10^{10}$ N/m², γ = 3500 kg/m³) then the ratio, $\sigma/\gamma = 0.05 \times 10^{-6}$ or $\sigma/\gamma = 20 \times 10^6$ (K = 2). Although (in 1976) the graphite fibers are strong ($\sigma/\gamma = 10 \times 10^6$), they are at best still ten times weaker than theory predicts.

Steel fiber has tensile strengths of 5,000 MPA (500 kg/mm^2), but the theoretic value is 22,000 MPa (1987). Polyethylene fiber has a tensile strength of 20,000 MPa and the theoretical value is 35,000 MPa (1987).

The mechanical behavior of nanotubes also has provided excitement because nanotubes are seen as the ultimate carbon fiber, which can be used as reinforcements in advanced composite technology. Early theoretical work and recent experiments on individual nanotubes (mostly MWNTs) have confirmed that nanotubes are one of the stiffest materials ever made. Whereas carbon–carbon covalent bonds are one of the strongest in nature, a structure based on a perfect arrangement of these bonds oriented along the axis of nanotubes would produce an exceedingly strong material. Traditional carbon fibers show high strength and stiffness, but fall far short of the theoretical in-plane strength of graphite layers (an order of magnitude lower). Nanotubes come close to being the best fiber that can be made from graphite structure.

For example, whiskers made from carbon nanotubes (CNT) have a tensile strength of 200 Giga-Pascals and Young's modulus of over 1 Tera Pascal (1999). The theory predicts 1 Tera Pascal and Young modulus 1–5 Tera Pascals. The hollow structure of nanotubes makes them very light (specific density varies from 0.8 g/cc for SWNTs up to 1.8 g/cc for MWNTs, compared to 2.26 g/cc for graphite or 7.8 g/cc for steel).

Specific strength (strength/density) is important in the design of our transportation system and space elevator; nanotubes have this value at least 2 orders of magnitude greater than steel. Traditional carbon fibers have a specific strength 40 times greater than steel. Where nanotubes are made of graphite carbon, they have good resistance to chemical attack and have high terminal stability. Oxidation studies have shown that the onset of oxidation shifts by about 100 °C higher temperatures in nanotubes compared to high modulus graphite fibers. In vacuums or reducing atmospheres, nanotubes structures will be stable at any practical service temperature. Nanotubes have excellent conductivity like copper.

The price for the SiC whiskers produced by Carborundun Co. with σ = 20,690 MPa, γ = 3.22 g/cc was $440/kg in 1989. Medicine, the environment, space, aviation, machine-building, and the computer industry need cheap nanotubes. Some American companies plan to produce nanotubes in 2–3 years.

Below the author provides a brief overview of the annual research information (2000) regarding the proposed experimental test fibers.

Data that can be used for computation

Let us consider the following experimental and industrial fibers, whiskers, and nanotubes:

1. Experimental nanotubes CNT (carbon nanotubes) have a tensile strength of 200 Giga-Pascals (20,000 kg/mm^2), Young's modulus is over 1 Tera Pascal, specific density γ=1800 kg/m^3 (1.8 g/cc) (year 2000). For safety factor n = 2.4, σ= 8300 kg/mm^2 = 8.3×10^{10} N/m^2, γ=1800 kg/m^3, (σ/γ)=46×10^6, K = 4.6. The SWNTs nanotubes have a density of 0.8 g/cc, and MWNTs have a density of 1.8 g/cc. Unfortunately, the nanotubes are very expensive at the present time (1994).

2. For whiskers C_D σ = 8000 kg/mm^2, γ = 3500 kg/m^3 (1989) [p.158][7].
3. For industrial fibers σ= 500 – 600 kg/mm^2, γ = 1800 kg/m^3, σ/γ = 2,78×10^6, K = 0.278 – 0.333,
Figures for some other experimental whiskers and industrial fibers are give in Table 1.2.

Table 1.2

Material Whiskers	Tensile strength kg/mm²	Density Fibers g/cc		MPa	Density g/cc
AlB_{12}	2650	2.6	QC-8805	6200	1.95
B	2500	2.3	TM9	6000	1.79
B_4C	2800	2.5	Thorael	5650	1.81
TiB_2	3370	4.5	Allien 1	5800	1.56
SiC	1380–4140	3.22	Allien 2	3000	0.97

See References [7, 8, 9, 10] in [1] Ch.1.

Conclusions

The new materials make the suggested transport system and projects highly realistic for a free trip to outer space without expention of energy. The same idea was used in the research and calculation of other revolutionary innovations such as launches into space without rockets (not space elevator, not gun); cheap delivery of loads from one continent to another across space; cheap delivery of fuel gas over long distances without steel tubes and damage to the environment; low cost delivery of large load flows across sea streams and mountains without bridges or underwater tunnels [Gibraltar, English Channel, Bering Stream (USA–Russia), Russia–Sakhalin–Japan, etc.]; new economical transportation systems; obtaining inexpensive energy from air streams at high altitudes; etc. some of these are in reference [1][12–21] Ch.1.

The author has developed innovations, estimations, and computations for the above mentioned problems. Even though these projects seem impossible for the current technology, the author is prepared to discuss the project details with serious organizations that have similar research and development goals.

1. Man in Space without Space Suite*

The author proposes and investigates his old idea - a living human in space without the encumbrance of a complex space suit. Only in this condition can biological humanity seriously attempt to colonize space because all planets of Solar system (except the Earth) do not have suitable atmospheres. Aside from the issue of temperature, a suitable partial pressure of oxygen is lacking. In this case the main problem is how to satiate human blood with oxygen and delete carbonic acid gas (carbon dioxide). The proposed system would enable a person to function in outer space without a space suit and, for a long time, without food. That is useful also in the Earth for sustaining working men in an otherwise deadly atmosphere laden with lethal particulates (in case of nuclear, chemical or biological war), in underground confined spaces without fresh air, under water or a top high mountains above a height that can sustain respiration.

---
* Published in [1] Ch.19; in [4] Ch.6.

Introduction

Short history. A fictional treatment of Man in space without spacesuit protection was famously treated by Arthur C. Clarke in at least two of his works, "Earthlight" and the more famous "2001: A Space Odyssey". In the scientific literature, the idea of sojourning in space without complex space suits was considered seriously about 1970 and an initial research was published in [1] p.335 – 336. Here is more detail research this possibility.

Humans and vacuum. Vacuum is primarily an asphyxiant. Humans exposed to vacuum will lose consciousness after a few seconds and die within minutes, but the symptoms are not nearly as graphic as commonly shown in pop culture. Robert Boyle was the first to show that vacuum is lethal to small animals. Blood and other body fluids do boil (the medical term for this condition is ebullism), and the vapour pressure may bloat the body to twice its normal size and slow circulation, but tissues are elastic and porous enough to prevent rupture. Ebullism is slowed by the pressure containment of blood vessels, so some blood remains liquid. Swelling and ebullism can be reduced by containment in a flight suit. Shuttle astronauts wear a fitted elastic garment called the Crew Altitude Protection Suit (CAPS) which prevents ebullism at pressures as low as 15 Torr (2 kPa). However, even if ebullism is prevented, simple evaporation of blood can cause decompression sickness and gas embolisms. Rapid evaporative cooling of the skin will create frost, particularly in the mouth, but this is not a significant hazard.

Animal experiments show that rapid and complete recovery is the norm for exposures of fewer than 90 seconds, while longer full-body exposures are fatal and resuscitation has never been successful.[4] There is only a limited amount of data available from human accidents, but it is consistent with animal data. Limbs may be exposed for much longer if breathing is not impaired. Rapid decompression can be much more dangerous than vacuum exposure itself. If the victim holds his breath during decompression, the delicate internal structures of the lungs can be ruptured, causing death. Eardrums may be ruptured by rapid decompression, soft tissues may bruise and seep blood, and the stress of shock will accelerate oxygen consumption leading to asphyxiation.

In 1942, the Nazi regime tortured Dachau concentration camp prisoners by exposing them to vacuum. This was an experiment for the benefit of the German Air Force (Luftwaffe), to determine the human body's capacity to survive high altitude conditions.

Some extremophile microrganisms, such as Tardigrades, can survive vacuum for a period of years.

Respiration (physiology). In animal physiology, respiration is the transport of oxygen from the clean air to the tissue cells and the transport of carbon dioxide in the opposite direction. This is in contrast to the biochemical definition of respiration, which refers to cellular respiration: the metabolic process by which an organism obtains energy by reacting oxygen with glucose to give water, carbon dioxide and ATP (energy). Although physiologic respiration is necessary to sustain cellular respiration and thus life in animals, the processes are distinct: cellular respiration takes place in individual cells of the animal, while physiologic respiration concerns the bulk flow and transport of metabolites between the organism and external environment.

In unicellular organisms, simple diffusion is sufficient for gas exchange: every cell is constantly bathed in the external environment, with only a short distance for gases to flow across. In contrast, complex multicellular organisms such as humans have a much greater distance between the environment and their innermost cells, thus, a respiratory system is needed for effective gas exchange. The respiratory system works in concert with a circulatory system to carry gases to and from the tissues.

In air-breathing vertebrates such as humans, respiration of oxygen includes four stages:

- *Ventilation* from the ambient air into the alveoli of the lung.
- *Pulmonary gas exchange* from the alveoli into the pulmonary capillaries.
- *Gas transport* from the pulmonary capillaries through the circulation to the peripheral capillaries in the organs.
- *Peripheral gas exchange* from the tissue capillaries into the cells and mitochondria.

Note that ventilation and gas transport require energy to power mechanical pumps (the diaphragm and heart respectively), in contrast to the passive diffusion taking place in the gas exchange steps.
Respiratory physiology is the branch of human physiology concerned with respiration.

Respiration system. In humans and other mammals, the respiratory system consists of the airways, the lungs, and the respiratory muscles that mediate the movement of air into and out of the body. Within the alveolar system of the lungs, molecules of oxygen and carbon dioxide are passively exchanged, by diffusion, between the gaseous environment and the blood. Thus, the respiratory system facilitates oxygenation of the blood with a concomitant removal of carbon dioxide and other gaseous metabolic wastes from the circulation. The system also helps to maintain the acid-base balance of the body through the efficient removal of carbon dioxide from the blood.

Circulation. The right side of the heart pumps blood from the right ventricle through the pulmonary semilunar valve into the pulmonary trunk. The trunk branches into right and left pulmonary arteries to the pulmonary blood vessels. The vessels generally accompany the airways and also undergo numerous branchings. Once the gas exchange process is complete in the pulmonary capillaries, blood is returned to the left side of the heart through four pulmonary veins, two from each side. The pulmonary circulation has a very low resistance, due to the short distance within the lungs, compared to the systemic circulation, and for this reason, all the pressures within the pulmonary blood vessels are normally low as compared to the pressure of the systemic circulation loop.
Virtually all the body's blood travels through the lungs every minute. The lungs add and remove many chemical messengers from the blood as it flows through pulmonary capillary bed. The fine capillaries also trap blood clots that have formed in systemic veins.

Gas exchange. The major function of the respiratory system is gas exchange. As gas exchange occurs, the acid-base balance of the body is maintained as part of homeostasis. If proper ventilation is not maintained, two opposing conditions could occur: 1) respiratory acidosis, a life threatening condition, and 2) respiratory alkalosis.
Upon inhalation, gas exchange occurs at the alveoli, the tiny sacs which are the basic functional component of the lungs. The alveolar walls are extremely thin (approx. 0.2 micrometres), and are permeable to gases. The alveoli are lined with pulmonary capillaries, the walls of which are also thin enough to permit gas exchange.

Membrane oxygenator. A membrane oxygenator is a device used to add oxygen to, and remove carbon dioxide from the blood. It can be used in two principal modes: to imitate the function of the lungs in cardiopulmonary bypass (CPB), and to oxygenate blood in longer term life support, termed Extracorporeal membrane oxygenation, ECMO. A membrane oxygenator consists of a thin gas permeable membrane separating the blood and gas flows in the CPB circuit; oxygen diffuses from the gas side into the blood, and carbon dioxide diffuses from the blood into the gas for disposal.
The introduction of microporous hollow fibres with very low resistance to mass transfer revolutionised design of membrane modules, as the limiting factor to oxygenator performance became the blood resistance [Gaylor, 1988]. Current designs of oxygenator typically use an extraluminal flow regime, where the blood flows outside the gas filled hollow fibres, for short term life support, while only the homogeneous membranes are approved for long term use.

Heart-lung machine. The heart-lung machine is a mechanical pump that maintains a patient's blood circulation and oxygenation during heart surgery by diverting blood from the venous system, directing it through tubing into an artificial lung (oxygenator), and returning it to the body. The oxygenator removes carbon dioxide and adds oxygen to the blood that is pumped into the arterial system.

Space suit. A space suit is a complex system of garments, equipment and environmental systems designed to keep a person alive and comfortable in the harsh environment of outer space. This applies to extra-vehicular activity (EVA) outside spacecraft orbiting Earth and has applied to walking, and riding the Lunar Rover, on the Moon.

Some of these requirements also apply to pressure suits worn for other specialized tasks, such as high-altitude reconnaissance flight. Above Armstrong's Line (~63,000 ft/~19,000 m), pressurized suits are needed in the sparse atmosphere. Hazmat suits that superficially resemble space suits are sometimes used when dealing with biological hazards.

A conventional space suit must perform several functions to allow its occupant to work safely and comfortably. It must provide: A stable internal pressure, Mobility, Breathable oxygen, Temperature regulation, Means to recharge and discharge gases and liquids , Means of collecting and containing solid and liquid waste, Means to maneuver, dock, release, and/or tether onto spacecraft.

Operating pressure. Generally, to supply enough oxygen for respiration, a spacesuit using pure oxygen must have a pressure of about 4.7 psi (32.4 kPa), equal to the 3 psi (20.7 kPa) partial pressure of oxygen in the Earth's atmosphere at sea level, plus 40 torr (5.3 kPa) CO_2 and 47 torr (6.3 kPa) water vapor pressure, both of which must be subtracted from the alveolar pressure to get alveolar oxygen partial pressure in 100% oxygen atmospheres, by the alveolar gas equation. The latter two figures add to 87 torr (11.6 kPa, 1.7 psi), which is why many modern spacesuits do not use 3 psi, but 4.7 psi (this is a slight overcorrection, as alveolar partial pressures at sea level are not a full 3 psi, but a bit less). In spacesuits that use 3 psi, the astronaut gets only 3 - 1.7 = 1.3 psi (9 kPa) of oxygen, which is about the alveolar oxygen partial pressure attained at an altitude of 6100 ft (1860 m) above sea level. This is about 78% of normal sea level pressure, about the same as pressure in a commercial passenger jet aircraft, and is the realistic lower limit for safe ordinary space suit pressurization which allows reasonable work capacity.

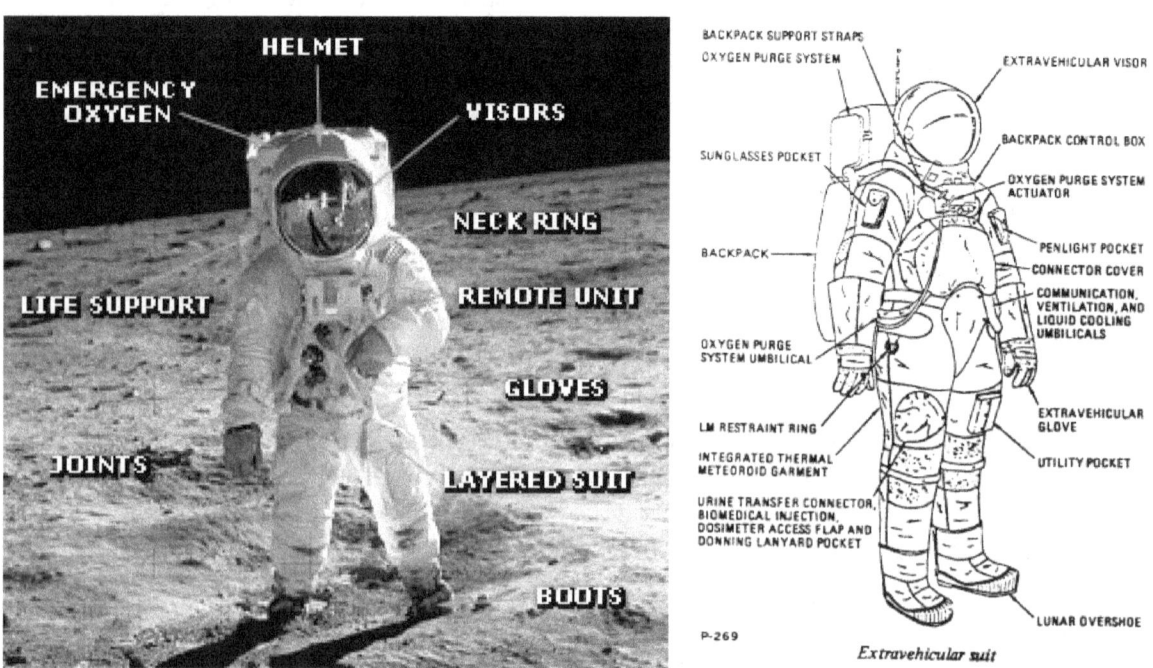

Fig. 2-1. a. Apollo 11 A7L space suit. b. Diagram showing component parts of A7L space suit.

Movements are seriously restricted in the suits, with a mass of more than 110 kilograms each (Shenzhou 7 space suit). The current space suits are very expensive. Flight-rated NASA spacesuits cost about $22,000,000. While other models may be cheaper, sale is not currently open even to the wealthy public. Even if spaceflight were free (a huge if) a person of average means could not afford to walk in space or upon other planets.

Brief Description of Innovation

A space suit is a very complex and expensive device (Fig. 2-1). Its function is to support the person's life, but it makes an astronaut immobile and slow, prevents him or her working, creates discomfort, does not allows eating in space, have a toilet, etc. Astronauts need a space ship or special space habitat located not far from away where they can undress for eating, toilet activities, and rest.

Why do we need a special space suit in outer space? There is only one reason – we need an oxygen atmosphere for breathing, respiration. Human evolution created lungs that aerates the blood with oxygen and remove carbon dioxide. However we can also do that using artificial apparatus. For example, doctors, performing surgery on someone's heart or lungs connect the patient to a heart – lung machine that acts in place of the patent's lungs or heart.

We can design a small device that will aerate the blood with oxygen and remove the carbon dioxide. If a tube from the main lung arteries could be connected to this device, we could turn on (off) the artificial breathing at any time and enable the person to breathe in a vacuum (on an asteroid or planet without atmosphere) in a degraded or poisonous atmosphere, or under water, for a long time. In space we can use a conventional Earth manufacture oversuit (reminiscent of those used by workers in semiconductor fabs) to protect us against solar ultraviolet light.

The sketch of device which saturates the blood with oxygen and removes the carbon dioxide is presented in fig.2-2. The Heart-Lung machines are widely used in current surgery.

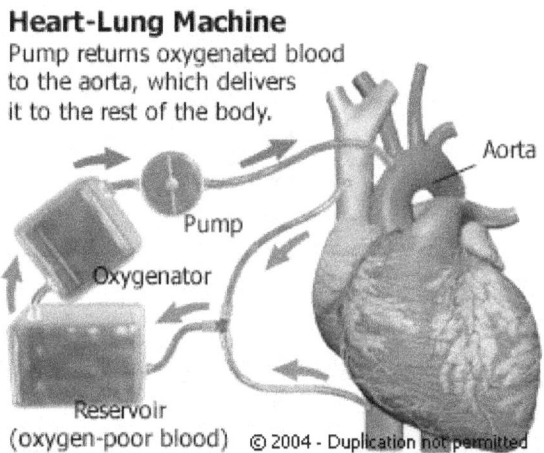

Fig.2-2. Principal sketch of heart-Lung Machine

The main part of this device is oxygenator, which aerates the blood with oxygen and removes the carbon dioxide. The principal sketch of typical oxygenator is presented in fig. 2-3.

The **circulatory system** is an organ system that moves nutrients, gases, and wastes to and from cells, helps fight diseases and helps stabilize body temperature and pH to maintain homeostasis. This system may be seen strictly as a blood distribution network, but some consider the circulatory system as composed of the cardiovascular system, which distributes blood, and the lymphatic system, which distributes lymph. While humans, as well as other vertebrates, have a closed cardiovascular system (meaning that the blood never leaves the network of arteries, veins and capillaries), some invertebrate groups have an open cardiovascular system. The most primitive animal phyla lack circulatory systems. The lymphatic system, on the other hand, is an open system.

The human blood circulatory system is shown in Fig. 2-5.

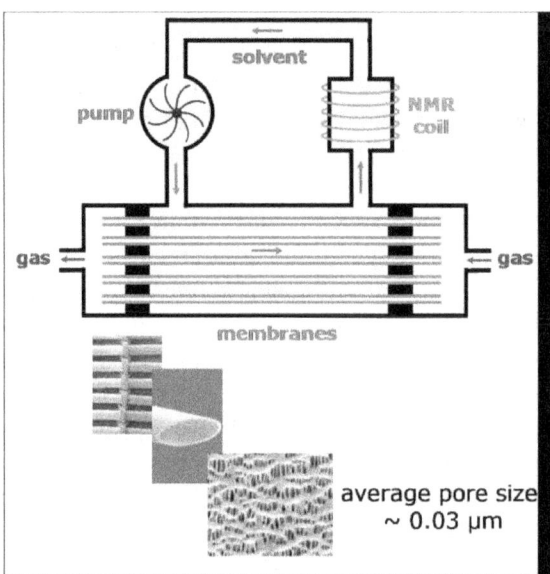

Fig.2-3. Principal sketch of oxygenator.

Current oxygenator is shown in Fig. 2-4.

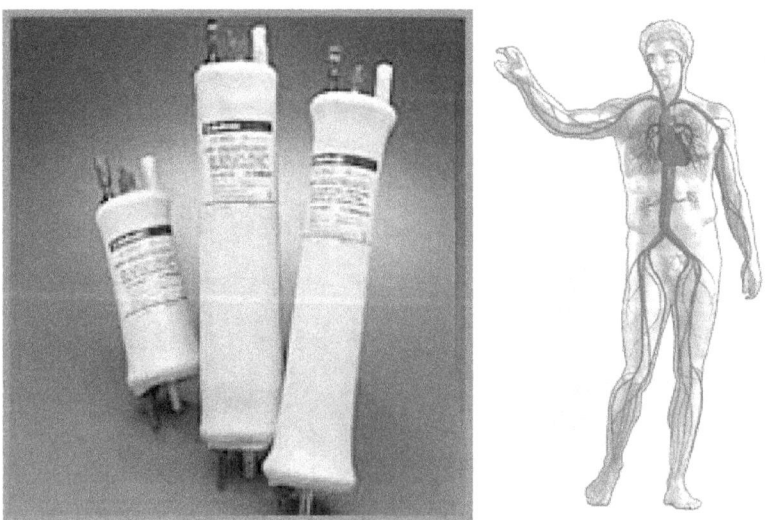

Fig. 2-4. (Left). Oxygenators.
Fig.2-5. (Right). The human circulatory system. Red indicates oxygenated blood, blue indicates deoxygenated.

The main components of the human circulatory system are the heart, the blood, and the blood vessels. The circulatory system includes: the pulmonary circulation, a "loop" through the lungs where blood is oxygenated; and the systemic circulation, a "loop" through the rest of the body to provide oxygenated blood. An average adult contains five to six quarts (roughly 4.7 to 5.7 liters) of blood, which consists of plasma that contains red blood cells, white blood cells, and platelets.

Two types of fluids move through the circulatory system: blood and lymph. The blood, heart, and blood vessels form the cardiovascular system. The lymph, lymph nodes, and lymph vessels form the lymphatic system. The cardiovascular system and the lymphatic system collectively make up the circulatory system.

The simplest form of intravenous access is a syringe with an attached **hollow needle.** The needle is inserted through the skin into a vein, and the contents of the syringe are injected through the needle into the bloodstream. This is most easily done with an arm vein, especially one of the metacarpal veins. Usually it is necessary to use a constricting band first to make the vein bulge; once the needle is in place, it is common to draw back slightly on the syringe to aspirate blood, thus verifying that the needle is really in a vein; then the constricting band is removed before injecting.

When man does not use the outer air pressure in conventional space suite, he not has opposed internal pressure except the heart small pressure in blood. The skin vapor easy stop by film clothes or make small by conventional clothes.

The current lung devices must be re-designed for space application. These must be small, light, cheap, easy in application (using hollow needles, no operation (surgery)!), work a long time in field conditions. Wide-ranging space colonization by biological humanity is impossible without them.

Artificial Nutrition.

Application of offered devices gives humanity a unique possibility to be a long time without conventional nutrition. Many will ask, "who would want to live like that?" But in fact many crew members, military, and other pressured personnel routinely cut short what most would consider normal dining routines. And there are those morbidly obese people for whom dieting is difficult exactly because (in an unfortunate phrase!) many can give up smoking 'cold turkey', but few can give up 'eating cold turkey'! Properly 'fed' intravenously, a person could lose any amount of excess weight he needed to, while not suffering hunger pains or the problems the conventional eating cycle causes. It is known that people in a coma may exist some years in artificial nutrition inserted into blood. Let us consider the current state of the art.

Total parenteral nutrition (TPN), is the practice of feeding a person intravenously, bypassing the usual process of eating and digestion. The person receives nutritional formulas containing salts, glucose, amino acids, lipids and added vitamins.

Total parenteral nutrition (TPN), also referred to as Parenteral nutrition (PN), is provided when the gastrointestinal tract is nonfunctional because of an interruption in its continuity or because its absorptive capacity is impaired. It has been used for comatose patients, although enteral feeding is usually preferable, and less prone to complications. Short-term TPN may be used if a person's digestive system has shut down (for instance by Peritonitis), and they are at a low enough weight to cause concerns about nutrition during an extended hospital stay. Long-term TPN is occasionally used to treat people suffering the extended consequences of an accident or surgery. Most controversially, TPN has extended the life of a small number of children born with nonexistent or severely deformed guts. The oldest were eight years old in 2003.

The preferred method of delivering TPN is with a medical infusion pump. A sterile bag of nutrient solution, between 500 mL and 4 L is provided. The pump infuses a small amount (0.1 to 10 mL/hr) continuously in order to keep the vein open. Feeding schedules vary, but one common regimen ramps up the nutrition over a few hours, levels off the rate for a few hours, and then ramps it down over a few more hours, in order to simulate a normal set of meal times.

Chronic TPN is performed through a central intravenous catheter, usually in the subclavian or jugular vein. Another common practice is to use a PICC line, which originates in the arm, and extends to one of the central veins, such as the subclavian. In infants, sometimes the umbilical vein is used.

Battery-powered ambulatory infusion pumps can be used with chronic TPN patients. Usually the pump and a small (100 ml) bag of nutrient (to keep the vein open) are carried in a small bag around the waist or on the shoulder. Outpatient TPN practices are still being refined.

Aside from their dependence on a pump, chronic TPN patients live quite normal lives.

Central IV lines flow through a catheter with its tip within a large vein, usually the superior vena cava or inferior vena cava, or within the right atrium of the heart.

There are several types of catheters that take a more direct route into central veins. These are collectively called *central venous lines*.

In the simplest type of central venous access, a catheter is inserted into a subclavian, internal jugular, or (less commonly) a femoral vein and advanced toward the heart until it reaches the superior vena cava or right atrium. Because all of these veins are larger than peripheral veins, central lines can deliver a higher volume of fluid and can have multiple lumens.

Another type of central line, called a Hickman line or Broviac catheter, is inserted into the target vein and then "tunneled" under the skin to emerge a short distance away. This reduces the risk of infection, since bacteria from the skin surface are not able to travel directly into the vein; these catheters are also made of materials that resist infection and clotting.

Testing

The offered idea may be easily investigated in animals on Earth by using currently available devices. The experiment includes the following stages:

1) Using a hollow needle, the main blood system of a good healthy animal connects to a current heart-lung machine.
2) The animal is inserted under a transparent dome and air is gradually changed to a neutral gas (for example, nitrogen). If all signs are OK, we may proceed to the following stage some days later.
3) The animal is inserted under a transparent dome and air is slowly (tens of minutes) pumped out. If all signs are OK we may start the following stage.
4) Investigate how long time the animal can be in vacuum? How quick we can decompress and compress? How long the animal may live on artificial nutrition? And so on.
5) Design the lung (oxygenator) devices for people which will be small, light, cheap, reliable, safe, which delete gases from blood (especially those that will cause 'bends' in the case of rapid decompression, and work on decreasing the decompressing time).
6) Testing the new devices on animals then human volunteers.

Advantages of offered system.

The offered method has large advantages in comparison with space suits:

1) The lung (oxygenator) devices are small, light, cheaper by tens to hundreds times than the current space suit.
2) It does not limit the activity of a working man.
3) The working time increases by some times. (less heat buildup, more supplies per a given carry weight, etc)
4) It may be widely used in the Earth for existing in poison atmospheres (industry, war), fire, rescue operation, under water, etc.
5) Method allows permanently testing (controlling) the blood and immediately to clean it from any poison and gases, wastes, and so on. That may save human lives in critical medical situations and in fact it may become standard emergency equipment.
6) For quick save the human life.
7) Pilots for high altitude flights.
8) The offered system is a perfect rescue system because you turn off from environment and exist INDEPENDENTLY from the environment. (Obviously excluding outside thermal effects, fires etc—but for

example, many fire deaths are really smoke inhalation deaths; the bodies are often not burned to any extent. In any case it is much easier to shield searing air from the lungs if you are not breathing it in!)

Conclusion.

The author proposes and investigates his old idea – a living human in space without the encumbrance of a complex space suit. Only in this condition can biological humanity seriously attempt to colonize space because all planets of Solar system (except the Earth) do not have suitable atmospheres. Aside from the issue of temperature, a suitable partial pressure of oxygen is lacking. In this case the main problem is how to satiate human blood with oxygen and delete carbonic acid gas (carbon dioxide). The proposed system would enable a person to function in outer space without a space suit and, for a long time, without food. That is useful also in the Earth for sustaining working men in an otherwise deadly atmosphere laden with lethal particulates (in case of nuclear, chemical or biological war), in underground confined spaces without fresh air, under water or a top high mountains above a height that can sustain respiration. There also could be numerous productive medical uses.

3.Electrostatic Levitation on Planet and Artificial Gravity for Space Ships and Asteroids*

The author offers and researches the conditions which allow people and vehicles to levitate on the Earth using the electrostatic repulsive force. He shows that by using small electrically charged balls, people and cars can take flight in the atmosphere. Also, a levitated train can attain high speeds. He has computed some projects and discusses the problems which can appear in the practical development of this method. It is also shown how this method may be used for creating artificial gravity (attraction force) into and out of space ships, space hotels, asteroids, and small planets which have little gravity. Unfortunately, the author is compelled to use the simplest calculations to evaluate various applications (projects).

------------------ -------------------------------

*Presented as paper AIAA-2005-4465 at 41 Propulsion Conference, 10–13 July 2005, Tucson, Arizona, USA. See also [1] Ch.15.

Introduction

People have dreamed about a flying freely in the air without any apparatus for many centuries. In ancient books we can find pictures of flying angels or God sitting on clouds or in heaven. At the present time you can see the same pictures on walls in many churches.

Physicist is know only two methods for creating repulsive force: magnetism and electrostatics. Magnetism is well studied and the use of superconductive magnets for levitating a train has been widely discussed in scientific journals, but repulsive magnets have only a short-range force. They work well for ground trains but are bad for air flight. Electrostatic flight needs powerful electric fields and powerful electric charges. The Earth's electric field is very weak and cannot be used for levitation. The main innovations presented in this chapter are methods for creating powerful static electrical fields in the atmosphere and powerful, stable electrical charges of small size which allow levitation (flight) of people, cars, and vehicles in the air. The author also shows how this method can be utilized into and out of a space ship (space hotel) or on an asteroid surface for creating artificial gravity. The author believes this method has applications in many fields of technology.

Magnetic levitation has been widely discussed in the literature for a long time. However, there are few scientific works related to electrostatic levitation. Electrostatic charges have a high voltage and can create corona discharges, breakthrough and relaxation. The Earth's electrostatic field is very weak and useless for flight. That is why many innovators think that electrostatic forces cannot be useful for levitation.

The author's first innovations in this field which changed this situation were offered in (1982)[1], and some practical applications were given in (1983)[1] Ch.15[2]. The idea was published in 1990 [1] Ch.15[3],p. 79[3]. In the following presented work, these ideas and innovations are researched in more detail. Some projects are also presented to allow estimation of the parameters of the offered flight systems.

Brief description of innovation

It is known that like electric charges repel, and unlike electric charges attract (Fig. 1a,b,c). A large electric charge (for example, positive) located at altitude induces the opposite (negative) electric charge at the Earth's surface (Figs. 1d,e,f,g) because the Earth is an electrical conductor. Between the upper and lower charges there is an electric field. If a small negative electric charge is placed in this electric field, this charge will be repelled from the like charges (on the Earth's surface) and attracted to the upper charge (Fig. 1d). That is the electrostatic lift force. The majority of the lift force is determined by the Earth's charges because the small charges are conventionally located near the Earth's surface. As shown below, these small charges can be connected to a man or a car and have enough force to lift and supports them in the air.

The upper charge may be located on a column as shown in Fig. 1d,e,f,g or a tethered air balloon (if we want to create levitation in a small town) (Fig. 1e), or air tube (if we want to build a big highway), or a tube suspended on columns (Fig.3-1f,g). In particular, the charges may be at two identically charged plates, used for a non-contact train (Fig.3- 3a).

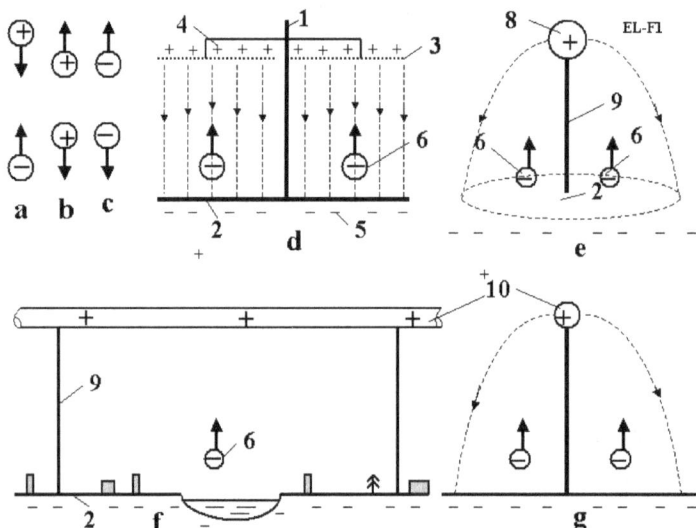

Fig. 3-1. Explanation of electrostatic levitation: a) Attraction of unlike charges; b,c) repulsion of like charges; d) Creation of the homogeneous electric field (highway); e) Electrical field from a large spherical charge ; f,g) Electrical field from a tube (highway) (side and front views). Notations are: 1, 9 – column, 2 – Earth (or other) surface charged by induction, 3 – net, 4 – upper charges, 5 – lower charges, 6 – levitation apparatus, 8 – charged air balloon, 9 – column, 10 – charged tube.

A lifting charge may use charged balls. If a thin film ball with maximum electrical intensity of below 3×10^6 V/m is used, the ball will have a radius of about 1 m (the man mass is 100 kg). For a 1 ton car, the ball will have a radius of about 3 m (see the computation below and Fig. 3-2g,h,i). If a higher electric intensity is used, the balls can be small and located underneath clothes (see below and Fig. 3-2 a,b,c).

The offered method has big advantages in comparison to conventional vehicles (Figs.3-1 and 3-2):

1) No very expensive highways are necessary. Rivers, lakes, forests, and buildings are not obstacles for this method.
2) In given regions (Figs. 3-1 and 3-2) people (and cars) can move at high speeds (man up 70 km/hour and cars up to 200–400 km/hour) in any direction using simple equipment (small balls under their clothing and small engines (Fig. 3- 2a,b,c)). They can perform vertical takeoffs and landings.
3) People can reduce their weight and move at high speed, jump a long distance, and lift heavy weights.
4) Building high altitude homes will be easier.

This method can be also used for a levitated train and artificial gravity in space ships, hotels, and asteroids (Fig. 3-3a,b)

A space ship (hotel) definitely needs artificial gravity. Any slight carelessness in space can result in the cosmonaut, instruments or devices drifting away from the space ship. Presently, they are connected to the space ship by cables, but this is not comfortable for working. Science knows only two methods of producing artificial gravity and attractive forces: rotation of space ship and magnetism. Both methods are bad. The rotation creates artificial gravity only inside the space ship (hotel). Observation of space from a rotating ship is very difficult. The magnetic force is only effective over a very short distance. The magnets stick together and a person has to expend a large effort to move (it is the same as when you are moving on a floor smeared with glue).

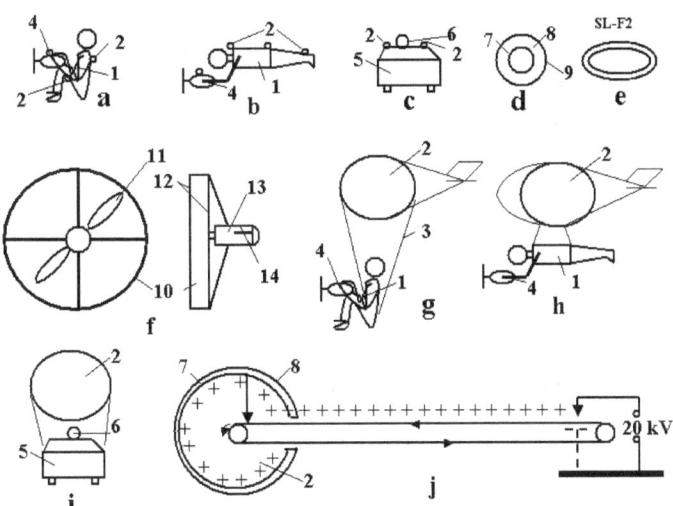

Fig. 3-2. Levitation apparatus: a,b) Single levitated man (mass up to 100 kg) using small highly charged balls 2. a) Sitting position; b) Reclining position; c) Small charged ball for levitating car; d) Small highly charged ball; e) Small highly charged cylindrical belt; f) Small air engine (forward and side views); g) Single levitated man (mass up to 100 kg) using a big non-highly charged ball which doesn't have an ionized zone (sitting position); h) The some man in a reclining position; i) Large charged ball to levitate a car which doesn't have an ionized zone; j) Installation for charging a ball using a Van de Graaff electrostatic generator (double generator potentially reaches 12 MV) in horizontal position.   Notations: 1 – man; 2 – charged lifting ball; 4 – handheld air engine; 5 – car; 6 – engine (turbo-rocket or other); 7 – conducting layer; 8 – insulator (dielectric); 9 – strong cover from

artificial fibers or whiskers; 10 – lagging; 11 – air propeller; 12 – preventive nets; 13 – engine; 14 – control knobs.

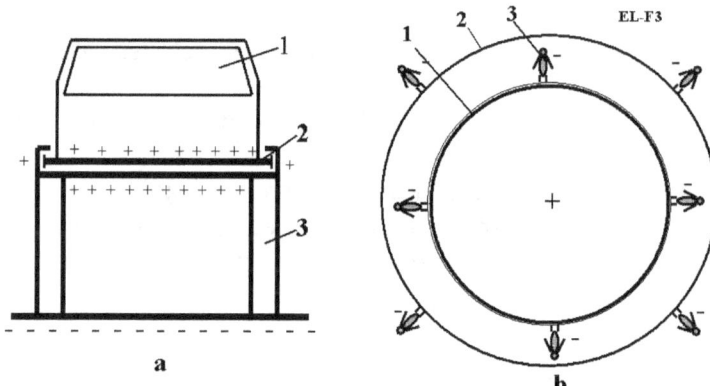

Fig. 3-3. Levitated train on Earth and artificial gravity into and on space ships and asteroids. a) Levitated train; b) Artificial gravity on a space ship. Notation: a) 1 – train; 2 – charged plates; 3 – insulated column; b) 1 – charged space body; 2 – space ship; 3 – man.

If then is a charge inside the space ship and small unlike charges attached to object elsewhere, then will fall back to the ship if they are dropped.

The same situation occurs for cosmonauts on asteroids or small planets which have very little gravity. If you charge the asteroid and cosmonauts with unlike electric charges, the cosmonauts will return to the asteroid during any walking and jumping.

The author acknowledges that this method has problems. For example, we need a high electrical intensity if we want to use small charged balls. This problem (and others) is discussed below.

Projects

Let us estimate the main parameters for some offered applications. Most people understand the magnitudes and properties of applications better than theoretical reasoning and equations. The suggested application parameters are not optimal, but our purpose is to show the method can be utilized by current technology.

1. Levitation Highway (Fig.1d).

The height of the top net is 20 m. The electrical intensity is $E_o = 2.5 \times 10^6$ V < $E_c = (3-4) \times 10^6$ V. The voltage between the top net and the ground is $U = 50 \times 10^6$ V. The width of each side of the road is 20 m. We first find the size of the lifting ball for the man (100 kg), car (1000 kg), or track (10,000 kg). Here R_c is the radius of the ionized zone [m]:

1) Flying man (mass M = 100 kg, ε = 3, E_i = 200×10^6 V/m, g ≈ 10 m/s^2)

$$E_a \leq \varepsilon E_i = 3 \times 200 \times 10^6, \quad a = \sqrt{\frac{kMg}{E_0 E_a}} = \sqrt{\frac{9 \times 10^9 \times 100 \times 10}{2,5 \times 10^6 \times 6 \times 10^8}} \approx 0.08 \, m, \quad R_c = \sqrt{\frac{E_a}{E_c}} \approx 1 \, m.$$

Notice that the radius of a single ball supporting the man is only 8 cm, or the man can use two balls a = 5–6 cm., R_c=0.75 m (or even more smaller balls). If the man uses a 1 m cylindrical belt, the radius of the belt cross-section area is 1.1 cm, σ = 100 kg/mm^2, E_a = 600×10^6 V/m ([1] Figs. 15.10 and 15.11). The belt may be more comfortable for some people.

2) With the same calculation you can find that a car of mass M = 1000 kg will be levitated using a single charged ball a = 23 cm, R_c = 3.2 m (or two balls with a = 16 cm. R_c = 2.3 m).

3) A truck of mass M = 10,000 kg will be levitated using a single charged ball a = 70 cm, R_c = 10 m (or two balls with a = 0.5 m. R_c = 7 m).

2. Levitating tube highway

Assume the levitation highway has the design of Fig. 1f,g where the top net is changed to a tube.

Take the data $E_0 = 2.5 \times 10^6$ V $< E_c = (3-4) \times 10^6$ V, $E_a = 2 \times 10^8$ V/m, $h = 20$ m. This means the electrical intensity, E_o, at ground level is the same as in the previous case. The required radius, a, of the top tube is

$$\frac{a}{h} = \frac{E_0}{2E_a} = 0.00625, \quad a = 0.00624h = 0.125 \, m, \quad R_c = a\frac{E_a}{E_0} = 10 \, m.$$

The diameter of the top tube is 0.25 m, the top ionized zone has a radius of 10 m.

3. Charged ball located on a high mast or tower

Assume there is a mast (tower) 500 m high with a ball of radius $a = 32$ m at its top charged up to $E_a = 3 \times 10^8$ V/m. The charge is

$$q = \frac{a^2 E_a}{k} = 34 \, C, \quad E_0 = k\frac{2q}{h^2} = 2.45 \times 10^6 \, V/m \, .$$

This electrical intensity at ground level means that within a radius of approximately 1 km, people, cars and other loads can levitate.

4. Levitation in low cumulonimbus and thunderstorm clouds

In these clouds the electrical intensity at ground level is about $E_0 = 3 \times 10^5 - 10^6$ V/m. A person can take more (or more highly charged) balls and levitate.

5. Artificial gravity on space ship's or asteroids

Assume the space ship is a sphere with an inner radius at $a = 10$ m and external radius of 13 m. We can create the electrical intensity $E_0 = 2.5 \times 10^6$ V/m without an ionized zone. The electrical charge is $q = a^2 E_0 / k = 2.8 \times 10^{-2}$ C. For a man weighing 100 kg (g = 10 m/s^2, force F = 1000 N), it is sufficient to have a charge of $q = F/E_0 = 4 \times 10^{-4}$ C and small ball with $a = 0.1$ m and $E_a = qk/a^2 = 3.6 \times 10^8$ V/m. In outer space at the ship's surface, the artificial gravity will be $(10/13)^2 = 0.6 = 60\%$[10] of g.

6. Charged ball as an accumulator of energy and rocket engine

The computations show the relative W/M energy calculated from safe tensile stress does not depend on E_a. A ball cover with a tensile stress of $\sigma = 200$ kg/mm^2 reaches 2.2 MJ/kg. This is close to the energy of conventional powder (3 MJ/kg). If whiskers or nanotubes are used the relative electrical storage energy will be close to than of liquid rocket fuel.

Two like charged balls repel one another and can give significant acceleration for a space vehicle, VTOL aircraft, or weapon.

Discussion

Electrostatic levitation could create a revolution in transportation, building, entertainment, aviation, space flights, and the energy industry.

The offered method needs development and testing. The experimental procedure it is not expensive. We just need a ball with a thin internal conducting layer, a dielectric cover, and high voltage charging equipment. This experiment can be carried out in any high voltage electric laboratory. The proposed levitation theory is based on proven electrostatic theory. There may be problems may be with discharging, blockage of the charge by the ionized zone, breakdown, and half-life of the discharge, but careful choice of suitable electrical materials and electric intensity may be also to solve them. Most of these problems do not occur in a vacuum.

Another problem is the affects of the strong electrostatic field on a living organism. Only experiments using animals can solve this. In any case, there are protection methods – conducting clothes or vehicle is (from metal or conducting paint) which offer a defense against the electric field.

4. A New Method of Atmospheric Reentry for Space Ships*

In recent years, industry has produced high-temperature fiber and whiskers. The author examined the atmospheric reentry of the USA Space Shuttles and proposed the use of high temperature tolerant parachute for atmospheric air braking. Though it is not large, a light parachute decreases Shuttle speed from 8 km/s to 1 km/s and Shuttle heat flow by 3 - 4 times. The parachute surface is opened with backside so that it can emit the heat radiation efficiently to Earth-atmosphere. The temperature of parachute is about 1000-1300°C. The carbon fiber is able to keep its functionality up to a temperature of 1500-2000°C. There is no conceivable problem to manufacture the parachute from carbon fiber. The proposed new method of braking may be applied to the old Space Shuttles as well as to newer spacecraft designs.

Introduction

In 1969 author applied a new method of global optimization to the problem of atmospheric reentry of spaceships ([2] Ch.8, Ref. [1] p. 188). The general analysis presented an additional method to the well-known method of outer space to Earth-atmosphere reentry ("high-speed corridor"). There is a low-speed corridor when the total heat is less than in a conventional high-speed passage. In that time for significantly decreasing the speed of a spaceship retro- and landing rocket engine needed to be used. That requires a lot of fuel. It is not acceptable for modern spaceships. Nowadays the textile industry produces heat resistant fiber that can be used for a new parachute system to be used in a high-temperature environment ([2] Ch.8, Ref.[2]-[4]).

The control is following: if $d\theta/dt > 0$ the all lift force $L = L_P = 0$. When the Shuttle riches the low speed the parachute area can be decreased or parachute can be detached. That case not computed. Used control is not optimal.

The results of integration are presented below. Used data: parachute area are S_P = 1000, 2000, 4000 m² (R_p = 17.8, 25.2, 35.7 m); m = 104,000 kg. The dash line is data of the Space Shuttle without a parachute.

Conclusion

The widespread production of high temperature fibers and whiskers allows us to design high-temperature tolerant parachutes, which may be used by space apparatus of all types for braking in a rarified planet atmosphere. The parachute has open backside surface that rapidly emits the heat radiation to outer space thereby quickly decreasing the parachute temperature. The proposed new method significantly decreases the maximum temperature and heat flow to main space apparatus. That decreases the heat protection mass and increases the useful load of the spacecraft. The method may be also used during an emergency reentering when spaceship heat protection is damaged (as in horrific instance of the Space Shuttle "Columbia").

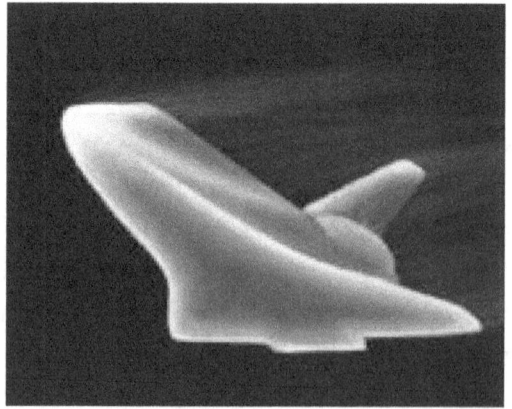

---

* Presented as Bolonkin's paper AIAA-2006-6985 to Multidisciplinary Analysis and Optimization Conference, 6-8 Sept. 2006, Portsmouth, Virginia. USA See also [2] Ch. 8.

Figure 4-1. Space Shuttle "Atlantic". during reentry.

Figure 4-2. The outside of the Shuttle heats to over 1,550 °C

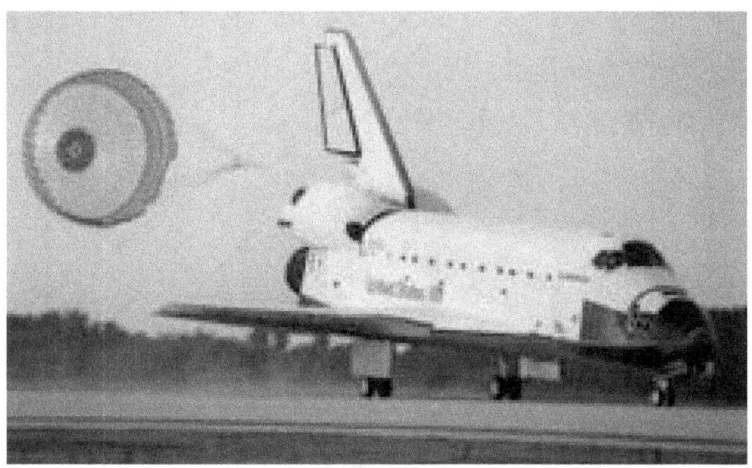

Figure 4-3. Endeavour deploys drag chute after touch-down.

Element*	Area, sq m (sq ft)	Weight, kg (lb)
FRSI	332.7 (3581)	532.1 (1173)
LRSI	254.6 (2741)	1014.2 (2236)
HRSI	479.7 (5164)	4412.6 (9728)
RCC	38.0 (409)	1697.3 (3742)
Miscellaneous		918.5 (2025)
Total	1105.0 (11895	8574.7 (18,904

*Includes bulk insulation, thermal barriers, and closeouts

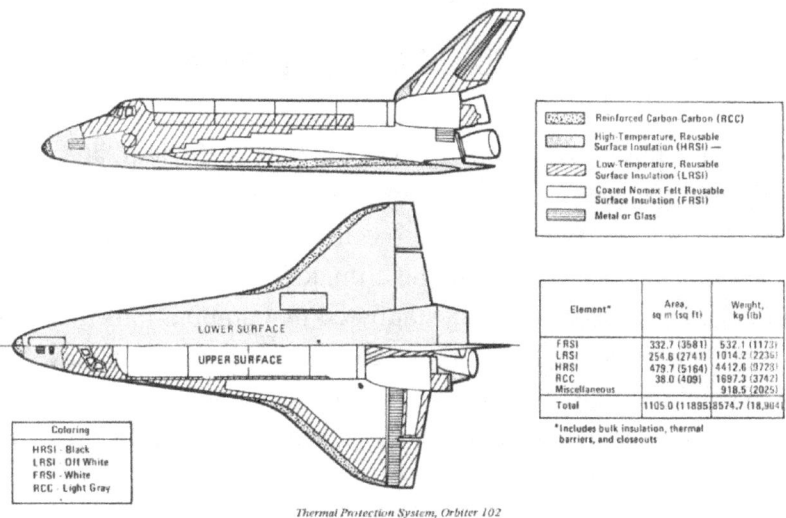

Figure 4-4. Space Shuttle Thermal Protection System Constituent Materials

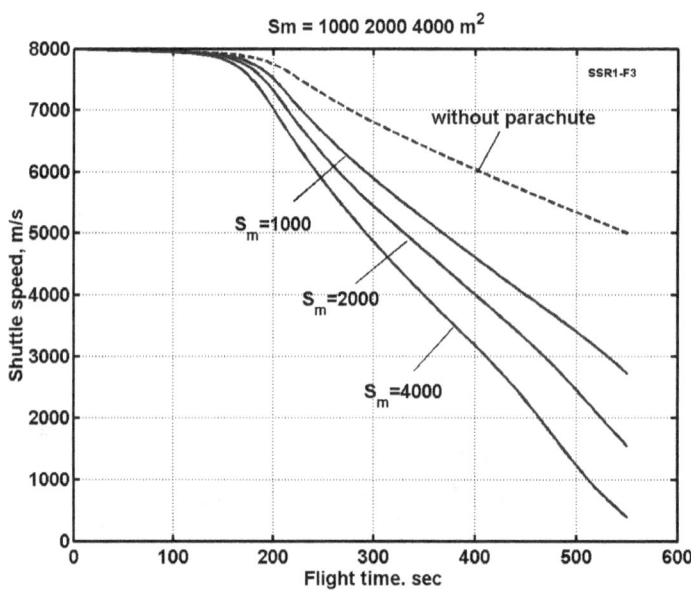

Figure 4-5. Decreasing of Space Shuttle speed with parachute and without it. $S_m = S_P$.

5. Inflatable Dome for Moon, Mars, Asteroids and Satellites[*]

On a planet without atmosphere, sustaining human life is very difficult and dangerous, especially during short sunlit period when low temperature prevails. To counter these environmental stresses, the author offer an innovative artificial "Evergreen" dome, an inflated hemisphere with interiors continuously providing a climate like that of Florida, Italy and Spain. The "Evergreen" dome theory is developed, substantiated by computations that show it is possible for current technology to construct and heat large enclosed volumes inexpensively. Specifically, a satisfactory result is reached by using high altitude magnetically supported sunlight reflectors and a special double thin film as an enclosing skin, which concentrates solar energy inside the dome while, at the same time, markedly decreasing the heat loss to exterior space. Offered design may be employed for settlements on the Moon, Mars, asteroids and satellites.

Introduction

The real development of outer space (permanent human life in space) requires two conditions: all-sufficient space settlement and artificial life conditions close to those prevailing currently on the Earth. (Such a goal extends what is already being attempted in the Earth-biosphere—for example at the 1st Advanced Architecture Contest, "Self-Sufficient Housing", sponsored by the Institute for Advanced Architecture of Catalonia, Spain, during 2006.) The first condition demands production of all main components needed for human life: food, oxidizer, and energy within the outer space and Solar System body colony.

The second requisite condition is a large surface settlement having useful plants, attractive flowers, splashing water pools, walking and sport areas, etc. All these conditions may be realized within large 'greenhouses' [1] that will produce food, oxidizer and "the good life" conditions.

---

[*] Presented as paper AIAA-2007-6262 to AIAA Conference "Space-2007", 18-20 September 2007, Long Beach. CA, USA. Published in http://arxiv.org. See also [3] Ch.2.

Human life in outer space and on other planetary or planet-like places will be more comfortable if it uses A.A. Bolonkin's macro-project proposal - staying in outer space without special spacesuit [2], p. 335 (mass of current spacesuit reaches 180 kg). The idea of this paper may be used also for control of Earth's regional and global weather and for converting our Earth's desert and cold polar zone regions into edenic subtropical gardens ([3] Ch.2, Ref.[3]-[4]).

The current conditions in Moon, Mars and Space are far from comfortable. For example, the Moon does not have any useful atmosphere, the day and night continues for 14 Earth days each, there are deadly space radiation and meteor bombardments, etc.

Especially during wintertime, Mars could provide only a meager and uncomfortable life-style for humans, offering low temperatures, strong winds. The distance north or south from that planet's equator is amongst the most significant measured environmental variables underlying the physical differences of the planet. In other words, future humans living in the Moon and Mars must be more comfortable for humans to explore and properly exploit these distant and dangerous places.

Possibly the first true architectural attempt at constructing effective artificial life-support systems on the climatically harsh Moon will be the building of greenhouses. Greenhouses are maintained nearly automatically by heating, cooling, irrigation, nutrition and plant disease management equipment. Humans share commonalities in their responses to natural environmental stresses that are stimulated by night cold, day heat, absent atmosphere, so on. Darkness everywhere inflicts the same personal visual discomfort and disorientation as cosmonauts/astronauts experience during their space-walks—that of being adrift in featureless space! With special clothing and shelters, humans can adapt successfully to the well-landmarked planet Mars, for example. Incontrovertibly, living on the Moon, beneath Mars' low-density atmosphere is difficult, even when tempered by strong conventional protective buildings.

Moon base.

'Evergreen' Inflated Domes

Our macro-engineering concept of inexpensive-to-construct-and-operate "Evergreen" inflated surface domes is supported by computations, making our macro-project speculation more than a daydream. Innovations are needed, and wanted, to realize such structures upon the Moon of our unique but continuously changing life.

Description and Innovations

Dome.

Our basic design for the Moon-Mars people-housing "Evergreen" dome is presented in Figure 5-1, which includes the thin inflated double film dome. The innovations are listed here: (1) the construction is air-inflatable; (2) each dome is fabricated with very thin, transparent film (thickness is 0.2 to 0.4 mm) without rigid supports; (3) the enclosing film is a two-layered structural element with air between the layers to provide insulation; (4) the construction form is that of a hemisphere, or in the instance of a roadway/railway a half-tube, and part of the film has control transparency and a thin aluminum layer about 1 μ or less that functions as the gigantic collector of incident solar radiation (heat). Surplus heat collected may be used to generate electricity or furnish mechanical energy; and (5) the dome is equipped with sunlight controlling louvers [also known as, "jalousie", a blind or shutter having adjustable slats to regulate the passage of air and sunlight] with one side thinly coated with reflective polished aluminum of about 1 μ thickness. Real-time control of the sunlight's entrance into the dome and nighttime heat's exit is governed by the shingle-like louvers or a controlled transparency of the dome film.

Variant 1 of artificial inflatable Dome for Moon and Mars is shown in Figure 5-1. Dome has top thin double film 4 covered given area and single under ground layer 6. The space between layers 4 - 6 is about 3 meters and it is filled by air. The support cables 5 connect the top and underground layers and Dome looks as a big air-inflated beach sunbathing or swimming mattress. The Dome includes hermetic sections connected by corridors 2 and hermetic lock chambers 3. Topmost film controls the dome's transparency (and reflectivity). That allows people to closely control temperature affecting those inside the dome. Topmost film also is of a double-thickness. When a meteorite pushes hole in the topmost double film, the lowermost layer closes the hole and puts temporary obstacles in the way of the escaping air. Dome has a fruitful soil layer, irrigation system, and cooling system 9 for supporting a selected given humidity. That is, a closed-biosphere with a closed life-cycle that regularly produces an oxidizer as well as sufficient food for people and their pets, even including some species of farm animals. Simultaneously, it is the beautiful and restful Earth-like place of abode. The offered design has a minimum specific mass, about 7-12 kg/m^2 (air - 3 kg, film - 1 kg, soil - 3 - 8 kg). Mass of an example area of 10×10 m is about 1 metric ton (oftentimes spelt "tonnes").

Figure 2 illustrates the second thin transparent dome cover we envision. The Dome has double film: semispherical layer (low pressure about 0.01 - 0.1 atmosphere, atm.) and lower layer (high 1 atm. pressure). The hemispherical inflated textile shell—technical "textiles" can be woven (weaving is an interlacement of warp and weft) or non-woven (homogenous films)—embodies the innovations listed: (1) the film is very thin, approximately 0.1 to 0.3 mm. A film this thin has never before been used in a major building; (2) the film has two strong nets, with a mesh of about 0.1 × 0.1 m and $a = 1 \times 1$ m, the threads are about 0.3 mm for a small mesh and about 1 mm for a big mesh.

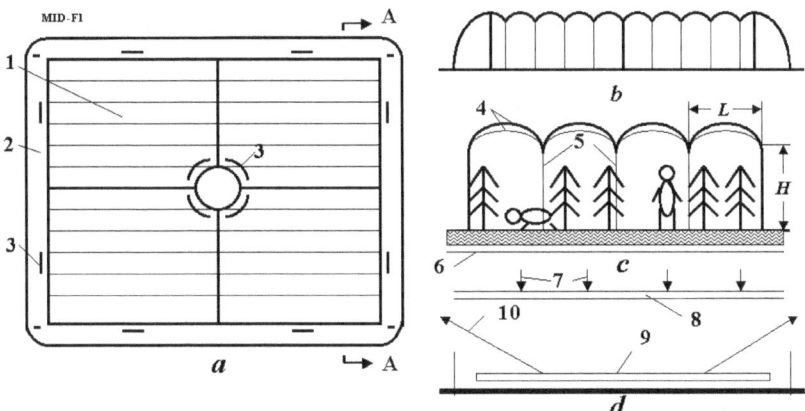

Figure 5-1. Variant 1 of artificial inflatable Dome for Moon and Mars. (a) top view of dome; (b) cross-section AA area of dome; (c) inside of the Dome; (d) Cooling system. Notations: 1 - internal section of Dome; 2 - passages; 3 - doors; 4 - transparence thin double film ("textiles") with control transparency; 5 - support cables; 6 - lower underground film; 7 - solar light; 8 - protection film; 9 - cooling tubes; 10 - radiation of cooling tubes.

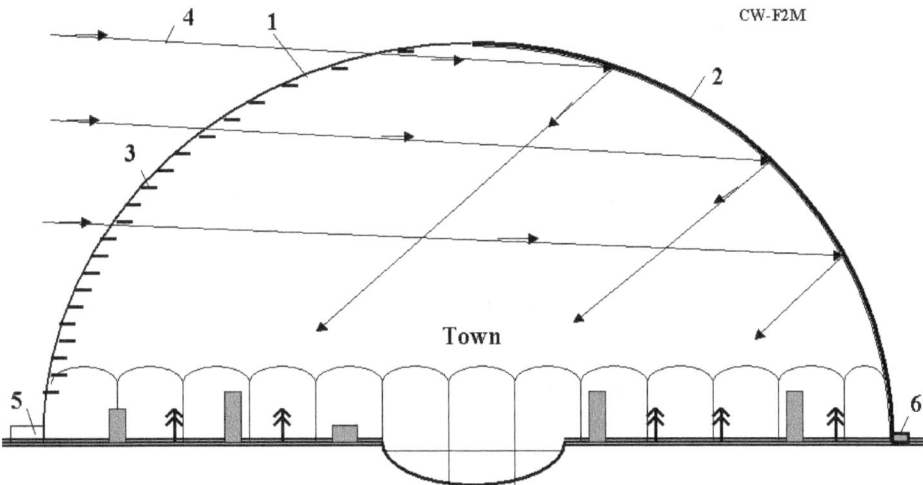

Figure 5-2. Variant 2 of artificial inflatable Dome for Moon and Mars. Notations: 1 - transparent thin double film ("textiles"); 2 - reflected cover of hemisphere; 3 - control louvers (jalousie); 4 - solar beams (light); 5 - enter (dock chamber); 6 - water extractor from air. The lower section has air pressure about 1 atm. The top section has air pressure of 0.01 - 0.1 atm.

The net prevents the watertight and airtight film covering from being damaged by micrometeorites; the film incorporates a tiny electrically-conductive wire net with a mesh of about 0.001 x 0.001 m and a line width of about 100 μ and a thickness near 1μ. The wire net can inform the "Evergreen" dome supervisors (human or automated equipment) concerning the place and size of film damage (tears, rips, punctures, gashes); the film is twin-layered with the gap — $c = 1$ m and $b = 2$ m—between the layer covering. This multi-layered covering is the main means for heat insulation and anti-puncture safety of a single layer because piercing won't cause a loss of shape since the film's second layer is unaffected by holing; the airspace in the dome's twin-layer covering can be partitioned, either hermetically or not; and part of the covering may have a very thin shiny aluminum coating that is about 1μ for reflection of non-useful or undesirable impinging solar radiation.

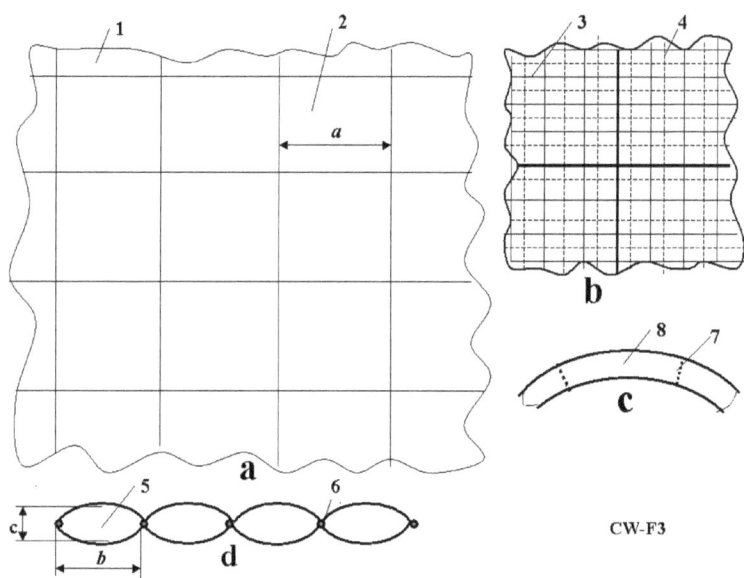

Figure 5-3. Design of "Evergreen" cover. Notations: (a) Big fragment of cover; (b) Small fragment of cover; (c) Cross-section of cover; (d) Longitudinal cross-section of cover; 1 - cover; 2 -mesh; 3 - small mesh; 4 - thin electric net; 5 - sell of cover; 6 - tubes; 7 - film partition (non hermetic); 8 - perpendicular cross-section area.

Offered inflatable Dome can cover a big region (town) and create beautiful Earth-like conditions on an outer space solid body (Figure 5-4a). In future, the "Evergreen" dome can cover a full planetary surface (Moon, Mars, asteroid) (Figure 5-4b). Same type of domes can cover the Earth's lands, converting them (desert, cool regions) into beautiful gardens with **controlled weather and closed** material life cycles.

Location, Illumination and Defending Human Settlements from Solar Wind and Space Radiation

The Moon makes one revolution in about 29 Earth days. If we want to have conventional Earth artificial day and natural solar lighting, the settlement must locate near one or both of the Moon's poles and have a magnetic control mirror suspended at high altitude in given (stationary) place (Figure 5-5). For building this mirror (reflector) may use idea and theory of magnetic levitation developed by A.A. Bolonkin in [3] Ch.2, Ref.[5]. If reflector is made with variable focus, as in [3] p. 306, Figure 16.3, then it may well be employed as a concentrator of sunlight and be harnessed for energy during "night" (Earth-time).

The second important feature of the offered installation is defense of the settlement from solar wind and all cosmic radiation. It is known that the Earth's magnetic field is a natural defense for living animals, plants and humans against high-energy particles, such as protons, of the solar wind. The artificial magnetic field near Moon settlement is hundreds of times stronger than the Earth's magnetic field. It will help to defend delicate humans. The polar location of the planned settlement also decreases the intensity of the solar wind. Location of human settlement in polar zone(s) Moon craters also decreases the solar wind radiation. People can move to an underground cosmic radiation protective shelter, a dugout or bunker, during periods of high Sun activity (solar flashes, coronal mass ejections).

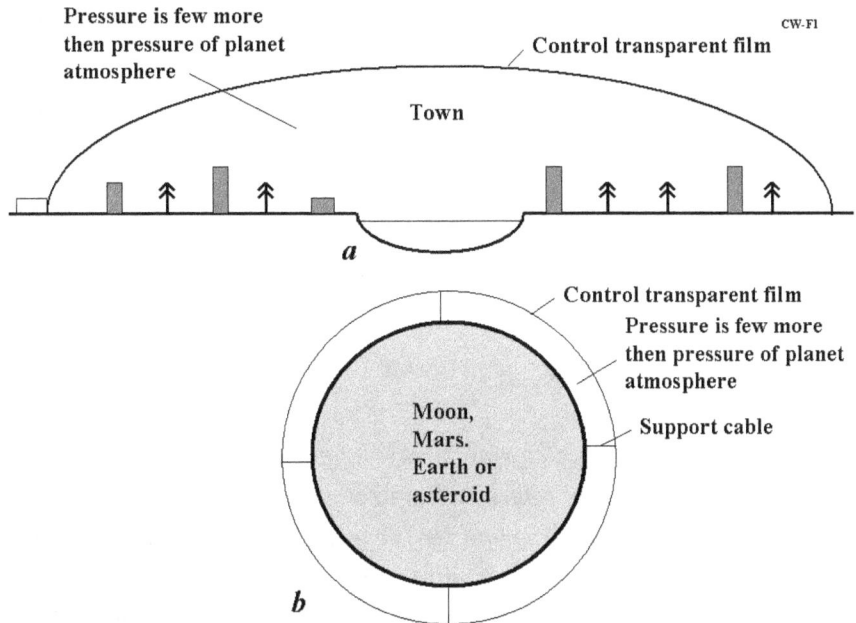

Figure 5-4. (*a*) Inflatable film dome over a single town; (*b*) Inflatable film dome covering a planet (Moon, Mars) and asteroid. Same type of domes can cover the Earth's extreme climate regions and convert them (desert, cool regions) into beautiful gardens with controlled weather and closed life cycles.

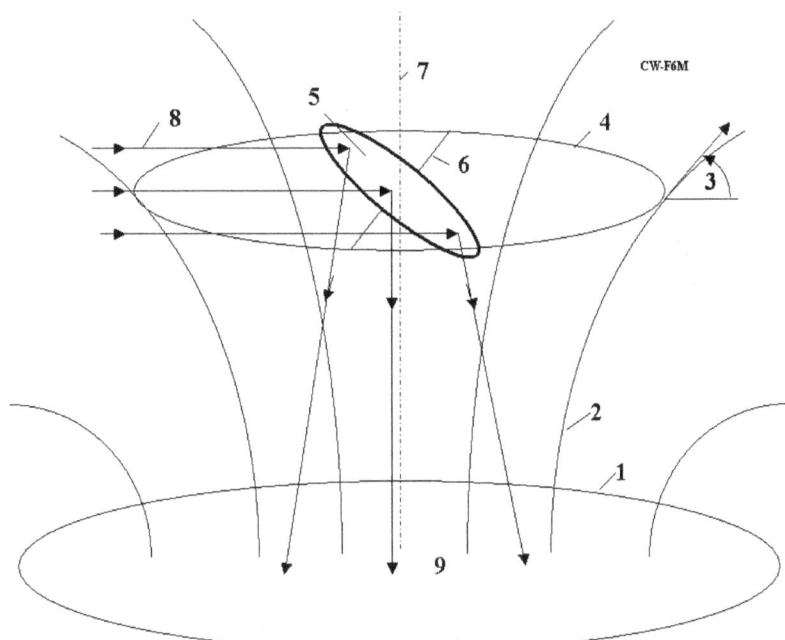

Figure 5-5. Magnetic control mirror is suspended at high altitude over human Moon settlement. Notations: 1 -superconductivity ground ring; 2 - magnetic lines of ground superconductivity ring; 3 -angle (α) between magnetic line of the superconductivity ground ring and horizontal plate (see Eq. (6)); top superconductivity ring for supporting the mirror (reflector) 5; 6 - axis of control reflector (which allows turning of mirror); 7 - vertical axis of the top superconductivity ring; 8 - solar light; 9 – human settlement.

The theory and computation of this installation is in theoretical section, below. The mass of the full reflector (rings, mirror, head screens is about 70 - 80 kg; if the reflector is used also as powerful energy source, then the mass can reach 100 - 120 kg. Note: for lifting, the reflector does not need a rocket. The magnetic force increases near ground (see Eq. (3)). This force lifts the reflector to the altitude that is required by its usage. The reflector also will be structurally stable because it is located in magnetic hole of a more powerful ground ring magnet.

The artificial magnetic field may be used, too, for free flying of men and vehicles, as it is described in [4]-[5]. If a planet does not have enough gravity, then electrostatic artificial gravity may be used [3], Ch. 15.

The magnetic force lifts the reflector to needed altitude.

Figure 6 illustrates a light-weight, possibly portable house, using the same basic construction materials as the dwelling/workplace.

Inflatable Space Hotel

We live during the 21st Century when Earthly polar tourism is just becoming a scheduled pastime and the world public anticipates outer space tourism. The offered inflatable outer space (satellite) hotel for tourists is shown in Figure 7. That has the common walking area (garden) covered by a film having the controlled transparency (reflectivity), internal sections (living rooms, offices, restaurants, concert hall, storage areas, etc.). Hotel has electrostatic artificial gravity [3], and magnetic field. The electrostatic artificial gravity creates usual Earth environment, the magnetic field allows people to easily fly near the outer space hotel and still be effectively defended from the dangerous, and sometimes even lethal, solar wind.

Hotel has electrostatic artificial gravity and magnetic field that will permit people to freely fly safely near the hotel even when radiation in outer space is closely present and intense.

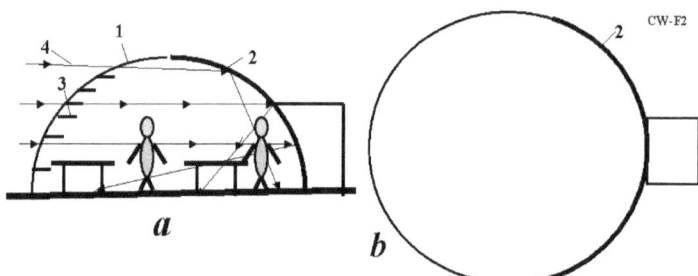

Figure 5-6. Inflatable film house for planet. Notation: (a) Cross-section area; (b) Top view. The other notations are same with Figure 5-2.

Visit Outer Space without a Spacesuit.

Current spacesuit designs are very complex and expensive "machines for living". They must, at minimum, unfailingly support human life for some period of time. However, the spacesuit makes a cosmonaut/astronaut barely mobile, slow moving, prevents exertive hard work, creates bodily discomfort such as pain or irritations, disallows meals in outer space, has no toilet, etc. Mass of current spacesuits is about 180 kg. Cosmonauts/Astronauts—these should be combined into "Spationauts" as the 20th Century descriptions were derived from Cold War superpower competition—must have spaceship or special outer space home habitat located not far from where they can undress for eating, toilet, and sleep as well as rest.

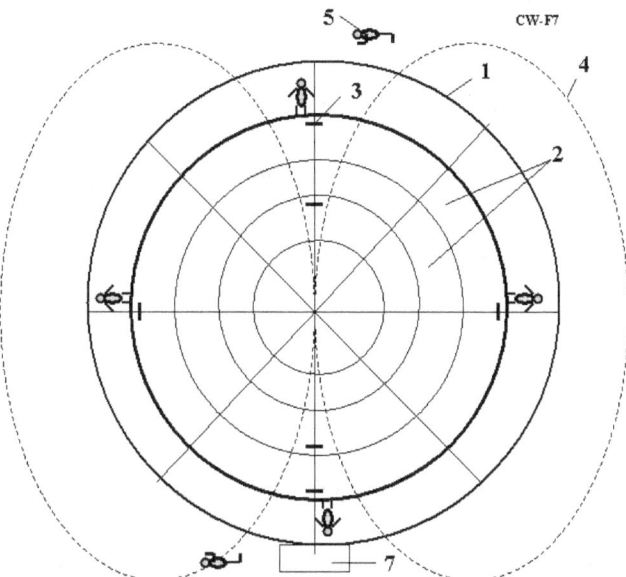

Figure 5-7. Inflatable space (satellite) hotel. Notations: 1 - inflatable hotel (control transparency cover film); 2 - internal sections of hotel (living rooms, offices, café, music hall, storage, etc.); 3 - door and windows in internal sections; 4 - magnetic line; 5 – outer space flying person (within hotel's magnetic field, [5]); 6 - common walking area (garden). 7 - docking chamber.

Why do humans need the special spacesuit in outer space, or on atmosphere-less bodies of the Solar System? There is only one reason – we need an oxygen atmosphere for breathing, respiration. Human evolution in the Earth-biosphere has created lungs that aerate our blood with oxygen and delete the carbonic acid. However, in a particularly harsh environment, we can do it more easily by artificial apparatus. For example, surgeons when they perform surgery on heart or lungs connect the patient to the apparatus "Heart-lung machine", temporarily stopping the patient's respiration and hear-beat. In [3] at p. 335, it is suggested that a method exists by donating some human blood, with the use of painless suture needles, is possible and that the blood can then be passed through artificial "lungs", just as is done in hospitals today.

We can design a small device that will aerate people's blood with oxygen infusion and delete the carbonic acid. To make offshoots from main lungs arteries to this device, we would turn on/off the artificial breathing at anytime and to be in vacuum (asteroid or planet without atmosphere) or bad or poisonous atmosphere, underwater a long time. In outer space we can be in conventional spacesuit defending the wearer from harmful solar light. Some type of girdle-like total body wrapping is required to keep persons in outer space from expanding explosively.

This idea may be checked with animal experiments in the Earth. We use the current "Heart-Lung" medical apparatus and put an animal under bell glass and remove the air inside the bell jar.

We can add into the blood all appropriate nutrition and, thusly, be without normal eating food for a long period of time; it is widely known that many humans in comas have lived fairly comfortably for many years entirely with artificial nourishment provided by drip injection.

The life possible in outer space without spacesuit will be easier, comfortable and entirely safe.

Macro-Projects

The dome shelter innovations outlined here can be practically applied to many cases and climatic regimes. We suggest initial macro-projects could be small (10 m diameter) houses (Figure 6) followed by an "Evergreen" dome covering a land area 200 m × 1000 m, with irrigated vegetation, homes, open-air swimming pools, playground, "under the stars style" concert hall.

The house and "Evergreen" dome have several innovations: magnetic suspended Sun reflector, double transparent insulating film, controllable jalousies coated with reflective aluminum (or film with transparency control properties and/or structures) and an electronic cable mesh inherent to the film for dome safety/integrity monitoring purposes. By undertaking to construct a half-sphere house, we can acquire experience in such constructions and explore more complex constructions. By computation, a 10 m diameter home has a useful floor area of 78.5 m^2, airy interior volume of 262 m^3 covered by an envelope with an exterior area of 157 m^2. Its film enclosure material would have a thickness of 0.0003 m with a total mass of about 100 kg.

A city-enclosing "Evergreen" dome of 200 m × 1000 m (Figure 2, with spherical end caps) could have calculated characteristics: useful area = 2.3×10^5 m^2, useful volume 17.8×10^6 m^3, exterior dome area of 3.75×10^5 m^2, comprised of a film of 0.0003 m thickness and about 200 tonnes. If the "Evergreen" dome were formed with concrete 0.25 m thick, the mass of the city-size envelope would be 200×10^3 tonnes, which is a thousand times heavier. Also, just for comparison, if we made a gigantic "Evergreen" dome with stiff glass, thousands of tonnes of steel, glass would be necessary and such materials would be very costly to transport hundreds or thousands of kilometers into outer space to the planet where they would be assembled by highly-paid but risk-taking construction workers. Our film is flexible and plastically deformable. It can be relatively cheap in terms of manufacturing cost. The single greatest boon to "Evergreen" dome construction, whether on the Moon, in Mars or elsewhere, is the protected cultivation of plants under a protective dome that efficiently captures energy from the available and technically harnessed sunlight.

Discussion

As with any innovative macro-project proposal, the reader will naturally have many questions. We offer brief answers to the two most obvious questions our readers are likely to ponder.

(1) *Cover damage.*

> The envelope contains a rip-stopping cable mesh so that the film cannot be damaged greatly. Its structured cross-section of double layering governs the escape of air inside the living realm. Electronic signals alert supervising personnel of all ruptures and permit a speedy repair effort by well-trained responsive emergency personnel. The topmost cover has a strong double film.

(2) *What is the design-life of the dome film covering?*

> Depending on the kinds of materials used, it may be as much a decade (up 30 years). In all or in part, the durable cover can be replaced periodically as a precautionary measure by its owners.

Conclusion

Utilization of "Evergreen" domes can foster the fuller economic development of the Moon, Mars and Earth itself - thus, increasing the effective area of territory dominated by humans on, at least, three celestial bodies. Normal human health can be maintained by ingestion of locally grown fresh vegetables and healthful "outdoor" exercise. "Evergreen" domes can also be used in the Earth's Tropics and Temperate Zone. Eventually, "Evergreen" domes may find application on the Moon or Mars since a vertical variant, inflatable space towers [1], are soon to become available for launching spacecraft inexpensively into Earth orbit or on to long-duration interplanetary outer spaceflights.

6. AB Method of Irrigation on planet without Water
(Closed-loop Water Cycle)*

Author methodically researched a revolutionary Macro-engineering idea for a closed-loop freshwater irrigation and in this chapter it is unveiled in some useful detail. We offer to cover a given site by a thin, enclosure film (with controlled heat conductivity and clarity) located at an altitude of 50 – 300 m. The film is supported, at its working altitude, by small additional induced air over-pressuring, and anchored to the ground by thin cables. We show that this closed dome allows full control of the weather within at a given planetary surface region (the day is always fine, it will rain only at night, no strong winds). The average Earth (white cloudy) reflectance equals 0.3 - 0.5. Consequently, Earth does lose about 0.3 - 0.5 of the maximum potential incoming solar energy. The dome (having control of the clarity of film and heat conductivity) converts the cold regions to controlled subtropics, hot deserts and desolate wildernesses to prosperous regions blessed temperate climate. This is, today, a realistic and cheap method of evaporation-economical irrigation and virtual weather control on Earth!

Description and Innovations

Our idea is a closed dome covering a local region by a thin film with controlled heat conductivity and optionally-controlled clarity (reflectivity, albedo, carrying capacity of solar spectrum)(Fig.6-1). The film is located at an altitude of ~50 – 300 m. The film is supported at this altitude by a small additional air pressure produced by ventilators sitting on the ground. The film is connected to Earth's surface by tethering cables. The cover may require double-layer film. We can control the heat conductivity of the dome cover by pumping in air between two layers of the dome film cover and change the solar heating due to sunlight heating by control of the cover's clarity. That allows selecting for different conditions (solar heating) in the covered area and by pumping air into dome. Envisioned is a cheap film having liquid crystal and conducting layers. The clarity is controlled by application of selected electric voltage. These layers, by selective control, can pass or blockade the available sunlight (or parts of solar spectrum) and pass or blockade the Earth's radiation. The incoming and outgoing radiations have different wavelengths. That makes control of them separately feasible and, therefore, possible to manage the heating or cooling of the Earth's surface under this film. In conventional conditions about 50% of the solar energy reaches the Earth surface. Much is reflected back to outer space by white clouds that shade approximately 65% of the Earth's land/water surface. In our closed water system the rain (or at least condensation) will occur at night when the temperature is low. In open atmosphere, the Sun heats the ground; the ground must heat the whole troposphere (4 – 5 km) before stable temperature rises happen. In our case, the ground heats ONLY the air in the dome (as in a hotbed). We have, then, a literal greenhouse effect. That means that many cold regions (Alaska, Siberia, northern Canada) may absorb more solar energy and became a temperate climate or sub-tropic climate (under the dome, as far as plants are concerned). That also means the Sahara and other deserts can be a prosperous regions with a fine growing and living climate and with a closed-loop water cycle.

The building of a film dome is easy. We spread out the film over Earth's surface, turn on the pumping propellers and the film is raised by air over-pressure to the needed altitude limited by the support cables. Damage to the film is not a major trouble because the additional air pressure is very small (0.0001- 0.01 atm) and air leakage is compensated for by spinning propeller pumps. Unlike in an outer space colony or extra-Earth planetary colony, the outside air is friendly and, at worst, we might lose some heat (or cold) and water vapor.

The first main innovation of our dome, and its main difference from a conventional hotbed, or glasshouse, is the inflatable HIGH span of the closed cover (up to 50 – 300 m). The elevated height of the enclosed volume

---
* Presented in electronic library of Cornel University http://arxiv.org in 27 December 2007. See also [3] Ch.1.

aids organizing of a CLOSED LOOP water cycle - accepting of water vaporized by plants and returning this water in the nighttime when the air temperature decreases. That allows us to perform irrigation in the vast area of Earth's land that does not have enough freshwater for agriculture. We can convert the desert and desolate wildernesses into Eden-like gardens without expensive delivery of remotely obtained and transported freshwater. The initial amount of freshwater for water cycle may be collected from atmospheric precipitation in some period or delivered. Prime soil is not a necessity because hydroponics allows us to achieve record harvests on any soil.

The second important innovation is using a cheap controlled heat conductivity, double-layer cover (controlled clarity is optionally needed for some regions). This innovation allows to conserve solar heat (in cold regions), to control temperature (in hot climates). That allows two to three rich crops annually in the Earth's middle latitudes and to conversion of the cold zones (Siberia, northern Canada, Alaska) to good single-crop regions.

The third innovation is control of the cover height, which allows adapting to local climatic seasons.

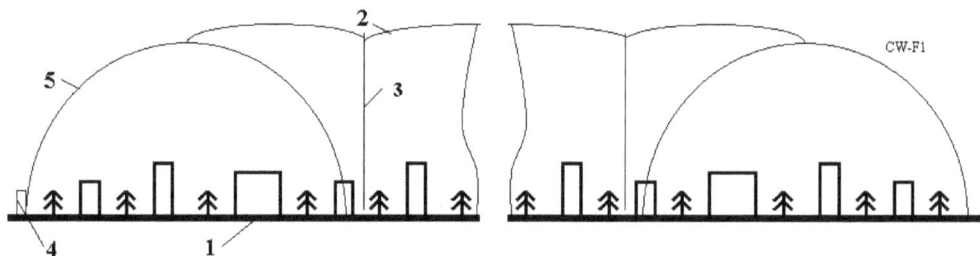

Figure 6-1. Film dome over agriculture region or a city. *Notations*: 1 - area, 2 - thin film cover with a control heat conductivity and clarity, 3 – control support cable and tubes for rain water (height is 50 – 300 m), 4 - exits and ventilators, 5 - semi-cylindrical border section.

The fourth innovation is the use of cheap, thin film as the high altitude cover. This innovation decreases the construction cost by thousands of times in comparison with the conventional very expensive glass-concrete domes offered by some for city use.

Lest it be objected that such domes would take impractical amounts of plastic, consider that the world's plastic production is today on the order of 100 million metric tons. If, with the expectation of future economic growth, this amount doubles over the next generation, and the increase is used for doming over territory at 500 tons a square kilometer, about 200,000 square kilometers could be roofed over annually. While small in comparison to the approximately 150 million square kilometers of land area, consider that 200,000 1 kilometer sites scattered over the face of the Earth made newly inhabitable could revitalize vast swaths of land surrounding them—one square kilometer could grow local vegetables for a city sited in the desert, one over there could grow bio-fuel, enabling a desolate South Atlantic island to become independent of costly fuel imports; at first, easily a billion people a year could be taken out of sweltering heat, biting cold and drenching rains, saving money that purchase, installation and operation of HVAC equipment—heating, ventilation, air-conditioning—would require.

Our dome design is presented in Figure 1 includes the thin inflated film dome. The innovations are listed here: (1) the construction is air-inflatable; (2) each dome is fabricated with very thin, transparent film (thickness is 0.1 to 0.3 mm) having controlled clarity and controlled heat conductivity without rigid supports; (3) the enclosing film has two conductivity layers plus a liquid crystal layer between them which changes its clarity, color and reflectivity under an electric voltage (Figure 6-2); (4) the bounded section of the dome proposed that has a hemispheric shape (#5, Figure 6-1). The air pressure is greater in these sections, and they protect the central sections from wind outside.

Figure 1 illustrates the thin transparent control dome cover we envision. The inflated textile shell—technical "textiles" can be woven or films—embodies the innovations listed: (1) the film is very thin, approximately 0.1 to 0.3 mm., implying under 500 tons per square kilometer. A film this thin has never before been used in a major building; (2) the film has two strong nets, with a mesh of about 0.1×0.1 m and $a = 1 \times 1$ m, the threads are about 0.5 mm for a small mesh and about 1 mm for a big mesh. The net prevents the watertight and airtight film covering from being damaged by vibration; (3) the film incorporates a tiny electrically conductive wire net with a mesh of about 0.1×0.1 m and a line width of about 100 μ and a thickness near 10 μ. The wire net is electric (voltage) control conductor. It can inform the dome maintenance engineers concerning the place and size of film damage (tears, rips); (4) the film may be twin-layered with the gap — $c = 1$ m and $b = 2$ m— between film layers for heat insulation. In Polar (and hot Tropic) regions this multi-layered covering is the main means for heat isolation and puncture of one of the layers wont cause a loss of shape because the second film layer is unaffected by holing; (5) the airspace in the dome's covering can be partitioned, either hermetically or not; and (6) part of the covering can have a very thin shiny aluminum coating that is about 1μ (micron) for reflection of unneeded sunlight in the equatorial region, or collect additional solar radiation in the polar regions [1].

The authors offer a method for moving off the accumulated snow and ice from the film in polar regions. After snowfall we decrease the heat cover protection, heating the snow (or ice) by warm air flowing into channels 5 (Figure 2) (between cover layers), and water runs down into tubes 3 (Figure 6-3).

The town cover may be used as a screen for projecting of pictures, films and advertising on the cover at nighttime.

Brief Data on Cover Film

Our dome filmic cover has 5 layers (Figure 4c): transparent dielectric layer, conducting layer (about 1 - 3 μ), liquid crystal layer (about 10 - 100 μ), conducting layer (for example, SnO_2), and transparent dielectric layer. Common thickness is 0.1 - 0.5 mm. Control voltage is 5 - 10 V. This film may be produced by industry relatively cheaply.

1. Liquid Crystals (LC)

Liquid crystals (LC) are substances that exhibit a phase of matter that has properties between those of a conventional liquid, and those of a solid crystal.

Liquid crystals find general employment in liquid crystal displays (LCD), which rely on the optical properties of certain liquid crystalline molecules in the presence or absence of an electric field. On command, the electric field can be used to make a pixel switch between clear or dark. Color LCD systems use the same technique, with color filters used to generate red, green, and blue pixels. Similar principles can be used to make other liquid crystal-based optical devices. Liquid crystal in fluid form is used to detect electrically generated hotspots for failure analysis in the semiconductor industry.

Liquid crystal memory units with extensive capacity were used in the USA's Space Shuttle navigation equipment. Worth noting also is the fact that many common fluids are, in fact, liquid crystals. Soap, for instance, is a liquid crystal, and forms a variety of LC phases depending on its concentration in water.

The conventional control clarity (transparancy) film reflected all superfluous energy to outer space. If the film has solar cells then it may convert the once superfluous solar energy into harnessed electricity.

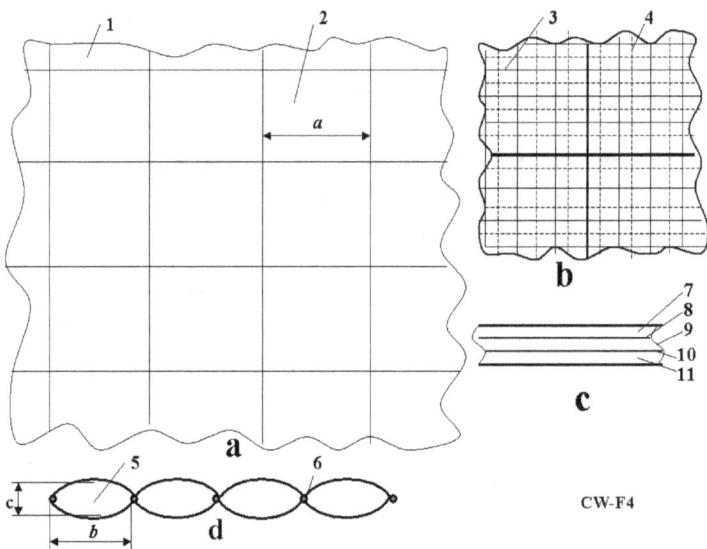

Figure 6-2. Design of membrane covering. *Notations*: (a) Big fragment of cover with control clarity (reflectivity, carrying capacity) and heat conductivity; (b) Small fragment of cover; (c) Cross-section of cover (film) having 5 layers; (d) Longitudinal cross-section of cover for cold and hot regions; 1 - cover; 2 -mesh; 3 - small mesh; 4 - thin electric net; 5 - cell of cover; 6 - tubes;: 7 - transparant dielectric layer, 8 - conducting layer (about 1 - 3 μ), 9 - liquid crystal layer (about 10 - 100 μ), 10 - conducting layer, and 11 - transparant dielectric layer. Common thickness is 0.1 - 0.5 mm. Control voltage is 5 - 10 V.

2. Transparency

In optics, transparency is the material property of passing natural and artificial light through any material. Though transparency usually refers to visible light in common usage, it may correctly be used to refer to any type of radiation. Examples of transparent materials are air and some other gases, liquids such as water, most non-tinted glasses, and plastics such as Perspex and Pyrex. The degree of material transparency varies according to the wavelength of the light. From electrodynamics it results that only a vacuum is really transparent in the strictist meaning, any matter has a certain absorption for electromagnetic waves. There are transparent glass walls that can be made opaque by the application of an electric charge, a technology known as electrochromics. Certain crystals are transparent because there are straight-lines through the crystal structure. Light passes almost unobstructed along these lines. There exists a very complicated scientific theory "predicting" (calculating) absorption and its spectral dependence of different materials.

3. Electrochromism

Electrochromism is the phenomenon displayed by some chemical species of reversibly changing color when a burst of electric charge is applied.

One good example of an electrochromic material is polyaniline which can be formed either by the electrochemical or chemical oxidation of aniline. If an electrode is immersed in hydrochloric acid which contains a small concentration of aniline, then a film of polyaniline can be grown on the electrode. Depending on the redox state, polyaniline can either be pale yellow or dark green/black. Other electrochromic materials that have found technological application include the viologens and polyoxotungstates. Other electrochromic materials include tungsten oxide (WO_3), which is the main chemical used in the production of electrochromic windows or smart windows.

As the color change is persistent and energy need only be applied to effect a change, electrochromic materials are used to control the amount of light and heat allowed to pass through windows ("smart windows"), and has also been applied in the automobile industry to automatically tint rear-view mirrors in various lighting conditions. Viologen is used in conjunction with titanium dioxide (TiO_2) in the creation of small digital displays. It is hoped that these will replace LCDs as the viologen (which is typically dark blue) has a high contrast to the bright color of the titanium white, therefore providing a high visibility of the display.

Conclusion

One half of Earth's human population is chronically malnourished. *The majority of Earth's surface area is not suitable for unshielded human life.* The increasing of agriculture area, crop capacity, carrying capacity by means of converting the deserts, desolate wildernesses, taiga, tundra permafrost into gardens are an important escape-hatch from some of humanity's most pressing macro-problems. The offered cheapest ($0.1÷ 0.3/m^2) AB method may dramatically increase the potentially realizable sown area, crop capacity; indeed the range of territory suitable for human living. In theory, converting all Earth land such as Alaska, northern Canada, Siberia, or the Sahara or Gobi deserts into prosperous gardens would be the equivalent of colonizing another Solar System planet. The suggested method is very cheap (cost of covering 1 m^2 is about 10 - 30 USA cents) and may be utilized immediately. We can start from small regions, such as towns in bad regions and, gradually, extend the practice over a large region—and what is as important, earning monetary profits most of the time.

Filmic domes can foster the fuller economic development of dry, hot, and cold regions such as the Earth's Arctic and Antarctic, the Sahara and, thus, increase the effective area of territory dominated by 21st Century humans. Normal human health can be maintained by ingestion of locally grown fresh vegetables and healthful "outdoor" exercise. The domes can also be used in the Tropics and Temperate Zone. Eventually, technical adaptations may find application on the Moon or Mars since a vertical variant, inflatable towers to outer space, are soon to become available for launching spacecraft inexpensively into Earth-orbit or interplanetary flights [12].

The related problems are researched in references [1]-[12].

Let us shortly summarize some advantages of this offered AB Dome method of climate moderation:

(1) Method does not need large amounts of constant input freshwater for irrigation;
(2) Low cost of inflatable filmic Dome per area reclaimed: (10 - 30 cents/m^2);
(3) Control of inside temperature is total;
(4) Usable in very hot and cool climate regions;
(5) Covered region is not at risk from exterior weather;
(6) Possibility of flourishing crops even with a sterile hydroponics soil;
(7) 2 – 3 harvests each year; without farmers' extreme normal risks.
(8) Rich harvests, at that.
(9) Converting deserts, desolate wildernesses, taiga, tundra, permafrost terrain, and the ocean into gardens;
(10) Covering towns and cities with low-cost, even picturesque domes;
(11) Using the dome cover for income-generating neighborhood illumination, picture displays, movies and videos as well as paid advertising.

We can concoct generally agreeable local weather and settle new territory for living with an agreeable climate (without daily rain, wind and low temperatures) for agriculture. By utilizing thin film, gigantic territorial expanses of dry and cold regions can be covered. Countries having big territory (but also much bad land) may be able to use domes to increase their population and became powerful states during the 21st Century.

The offered method may be used to conserve a vanishing sea such as the Aral Sea or the Dead Sea. A closed-loop water cycle could save these two seas for a future generation of people, instead of bequeathing a salty dustbowl.

A.A. Bolonkin has further developed the same method for the ocean. By controlling the dynamics and climate there, oceanic colonies may increase the Earth's humanly useful region another three times concurrent with or after the doubling of useful land outlined above. Our outlined method would allow the Earth's human population to increase by 5 – 10 times, without the starvation.

The offered method can solve the problem of apparent problem of global warming because the AB domes will be able to confine until use much carbon dioxide gas which appreciably increases a harvest. This carbon dioxide gas will show up in yet more productive crop harvests! The dome lift-force reaches up to 300 kg/m^2. The telephone, TV, electric, water and other communications can be suspended from the dome cover.

The offered method can also help to defend cities (or an entire given region) from rockets, nuclear warheads, and military aviation. Details are offered in a later chapter. This method may be applied to other planets and satellites as Moon and Mars.

7. Artificial Explosion of Sun and AB-Criterion for Solar Detonation*

The Sun contains ~74% hydrogen by weight. The isotope hydrogen-1 (99.985% of hydrogen in nature) is a usable fuel for fusion thermonuclear reactions.

This reaction runs slowly within the Sun because its temperature is low (relative to the needs of nuclear reactions). If we create higher temperature and density in a limited region of the solar interior, we may be able to produce self-supporting detonation thermonuclear reactions that spread to the full solar volume. This is analogous to the triggering mechanisms in a thermonuclear bomb. Conditions within the bomb can be optimized in a small area to initiate ignition, then spread to a larger area, allowing producing a hydrogen bomb of any power. In the case of the Sun certain targeting practices may greatly increase the chances of an artificial explosion of the Sun. This explosion would annihilate the Earth and the Solar System, as we know them today.

The reader naturally asks: Why even contemplate such a horrible scenario? It is necessary because as thermonuclear and space technology spreads to even the least powerful nations in the centuries ahead, a dying dictator having thermonuclear missile weapons can produce (with some considerable mobilization of his military/industrial complex)— an artificial explosion of the Sun and take into his grave the whole of humanity. It might take tens of thousands of people to make and launch the hardware, but only a very few need know the final targeting data of what might be otherwise a weapon purely thought of (within the dictator's defense industry) as being built for peaceful, deterrent use.

Those concerned about Man's future must know about this possibility and create some protective system—or ascertain on theoretical grounds that it is entirely impossible.

Humanity has fears, justified to greater or lesser degrees, about asteroids, warming of Earthly climate, extinctions, etc. which have very small probability. But all these would leave survivors --nobody thinks that the terrible annihilation of the Solar System would leave a single person alive. That explosion appears possible at the present time. In this paper is derived the 'AB-Criterion' which shows conditions wherein the artificial explosion of Sun is possible. The author urges detailed investigation and proving or disproving of this rather horrifying possibility, so that it may be dismissed from mind—or defended against.

----

* This work is written together J. Friedlander. He corrected the author's English, wrote together with author Abstract, Sections 8, 10 ("Penetration into Sun" and "Results"), and wrote Section 11 "Discussion" as the solo author. See also [4] Ch.10.

Statement of Problem, Main Idea and Our Aim

The present solar temperature is far lower than needed for propagating a runaway thermonuclear reaction. In Sun core the temperature is only ~13.6 MK (0.0012 MeV). The Coulomb barrier for protons (hydrogen) is more then 0.4 MeV. Only very small proportions of core protons take part in the thermonuclear reaction (they use a tunnelling effect). Their energy is in balance with energy emitted by Sun for the Sun surface temperature 5785 K (0.5 eV).

We want to clarify: If we create a zone of limited size with a high temperature capable of overcoming the Coulomb barrier (for example by insertion of a thermonuclear warhead) into the solar photosphere (or lower), can this zone ignite the Sun's photosphere (ignite the Sun's full load of thermonuclear fuel)? Can this zone self-support progressive runaway reaction propagation for a significant proportion of the available thermonuclear fuel?

If it is possible, researchers can investigate the problems: What will be the new solar temperature? Will this be metastable, decay or runaway? How long will the transformed Sun live, if only a minor change? What the conditions will be on the Earth?

Why is this needed?

As thermonuclear and space technology spreads to even the least powerful nations in the decades and centuries ahead, a dying dictator having thermonuclear weapons and space launchers can produce (with some considerable mobilization of his military/industrial complex)— the artificial explosion of the Sun and take into his grave the whole of humanity.

It might take tens of thousands of people to make and launch the hardware, but only a very few need know the final targeting data of what might be otherwise a weapon purely thought of (within the dictator's defense industry) as being built for peaceful, 'business as usual' deterrent use. Given the hideous history of dictators in the twentieth century and their ability to kill technicians who had outlived their use (as well as major sections of entire populations also no longer deemed useful) we may assume that such ruthlessness is possible.

Given the spread of suicide warfare and self-immolation as a desired value in many states, (in several cultures—think Berlin or Tokyo 1945, New York 2001, Tamil regions of Sri Lanka 2006) what might obtain a century hence? All that is needed is a supportive, obedient defense complex, a 'romantic' conception of mass death as an ideal—even a religious ideal—and the realization that his own days at power are at a likely end. It might even be launched as a trump card in some (to us) crazy internal power struggle, and plunged into the Sun and detonated in a mood of spite by the losing side. *'Burn baby burn'!*

A small increase of the average Earth's temperature over 0.4 K in the course of a century created a panic in humanity over the future temperature of the Earth, resulting in the Kyoto Protocol. Some stars with active thermonuclear reactions have temperatures of up to 30,000 K. If not an explosion but an enchanced burn results the Sun might radically increase in luminosity for –say--a few hundred years. This would suffice for an average Earth temperature of hundreds of degrees over 0 C. The oceans would evaporate and Earth would bake in a Venus like greenhouse, or even lose its' atmosphere entirely.

Thus we must study this problem to find methods of defense from human induced Armageddon.

The interested reader may find needed information in [4] Ch.10, Ref. [1]-[4].

Results of research

The Sun contains 73.46 % hydrogen by weight. The isotope hydrogen-1 (99.985% of hydrogen in nature) is usable fuel for a fusion thermonuclear reaction.

The p-p reaction runs slowly within the Sun because its temperature is low (relative to the temperatures of nuclear reactions). If we create higher temperature and density in a limited region of the solar interior, we may be able to produce self-supporting, more rapid detonation thermonuclear reactions that may spread to the full solar volume. This is analogous to the triggering mechanisms in a thermonuclear bomb. Conditions within the bomb can be optimized in a small area to initiate ignition, build a spreading reaction and then feed it into a larger area, allowing producing a 'solar hydrogen bomb' of any power—but not necessarily one whose power can be limited. In the case of the Sun certain targeting practices may greatly increase the chances of an artificial explosion of the entire Sun. This explosion would annihilate the Earth and the Solar System, as we know them today.

Author A.A. Bolonkin has researched this problem and shown that an artificial explosion of Sun cannot be precluded. In the Sun's case this lacks only an initial fuse, which induces the self-supporting detonation wave. This research has shown that a thermonuclear bomb exploded within the solar photosphere surface may be the fuse for an accelerated series of hydrogen fusion reactions.

The temperature and pressure in this solar plasma may achieve a temperature that rises to billions of degrees in which all thermonuclear reactions are accelerated by many thousands of times. This power output would further heat the solar plasma. Further increasing of the plasma temperature would, in the worst case, climax in a solar explosion.

The possibility of initial ignition of the Sun significantly increases if the thermonuclear bomb is exploded under the solar photosphere surface. The incoming bomb has a diving speed near the Sun of about 617 km/sec. Warhead protection to various depths may be feasible –ablative cooling which evaporates and protects the warhead some minutes from the solar temperatures. The deeper the penetration before detonation the temperature and density achieved greatly increase the probability of beginning thermonuclear reactions which can achieve explosive breakout from the current stable solar condition.

Compared to actually penetrating the solar interior, the flight of the bomb to the Sun, (with current technology requiring a gravity assist flyby of Jupiter to cancel the solar orbit velocity) will be easy to shield from both radiation and heating and melting. Numerous authors, including A.A. Bolonkin in works [7]-[12] offered and showed the high reflectivity mirrors which can protect the flight article within the orbit of Mercury down to the solar surface.

The author A.A. Bolonkin originated the AB Criterion, which allows estimating the condition required for the artificial explosion of the Sun.

Discussion

If we (humanity—unfortunately in this context, an insane dictator representing humanity for us) create a zone of limited size with a high temperature capable of overcoming the Coulomb barrier (for example by insertion of a specialized thermonuclear warhead) into the solar photosphere (or lower), can this zone ignite the Sun's photosphere (ignite the Sun's full load of thermonuclear fuel)? Can this zone self-support progressive runaway reaction propagation for a significant proportion of the available thermonuclear fuel?

If it is possible, researchers can investigate the problems: What will be the new solar temperature? Will this be metastable, decay or runaway? How long will the transformed Sun live, if only a minor change? What the conditions will be on the Earth during the interval, if only temporary? If not an explosion but an enhanced burn results the Sun might radically increase in luminosity for –say--a few hundred years. This would suffice for an average Earth temperature of hundreds of degrees over 0 °C. The oceans would evaporate and Earth would bake in a Venus like greenhouse, or even lose its' atmosphere entirely.

It would not take a full scale solar explosion, to annihilate the Earth as a planet for Man. (For a classic report on what makes a planet habitable, co-authored by Issac Asimov, see http://www.rand.org/pubs/commercial_books/2007/RAND_CB179-1.pdf 0.

Converting the sun even temporarily into a 'superflare' star, (which may hugely vary its output by many percent, even many times) over very short intervals, not merely in heat but in powerful bursts of shorter wavelengths) could kill by many ways, notably ozone depletion—thermal stress and atmospheric changes and hundreds of others of possible scenarios—in many of them, human civilization would be annihilated. And in many more, humanity as a species would come to an end.

The reader naturally asks: Why even contemplate such a horrible scenario? It is necessary because as thermonuclear and space technology spreads to even the least powerful nations in the centuries ahead, a dying dictator having thermonuclear missile weapons can produce (with some considerable mobilization of his military/industrial complex)— the artificial explosion of the Sun and take into his grave the whole of humanity. It might take tens of thousands of people to make and launch the hardware, but only a very few need know the final targeting data of what might be otherwise a weapon purely thought of (within the dictator's defense industry) as being built for peaceful, deterrent use.

Those concerned about Man's future must know about this possibility and create some protective system— or ascertain on theoretical grounds that it is entirely impossible, which would be comforting.

Suppose, however that some variation of the following is possible, as determined by other researchers with access to good supercomputer simulation teams. What, then is to be done?

The action proposed depends on what is shown to be possible.

Suppose that no such reaction is possible—it dampens out unnoticeably in the solar background, just as no fission bomb triggered fusion of the deuterium in the oceans proved to be possible in the Bikini test of 1946. This would be the happiest outcome.

Suppose that an irruption of the Sun's upper layers enough to cause something operationally similar to a targeted 'coronal mass ejection' – CME-- of huge size targeted at Earth or another planet? Such a CME like weapon could have the effect of a huge electromagnetic pulse. Those interested should look up data on the 1859 solar superstorm, the Carrington event, and the Stewart Super Flare. Such a CME/EMP weapon might target one hemisphere while leaving the other intact as the world turns. Such a disaster could be surpassed by another step up the escalation ladder-- by a huge hemisphere killing thermal event of ~12 hours duration such as postulated by science fiction writer Larry Niven in his 1971 story "Inconstant Moon"—apparently based on the Thomas Gold theory (ca. 1969-70) of rare solar superflares of 100 times normal luminosity. Subsequent research[18] (Wdowczyk and Wolfendale, 1977) postulated horrific levels of solar activity, ozone depletion and other such consequences might cause mass extinctions. Such an improbable event might not occur naturally, but could it be triggered by an interested party? A triplet of satellites monitoring at all times both the sun from Earth orbit and the 'far side' of the Sun from Earth would be a good investment both scientifically and for purposes of making sure no 'creative' souls were conducting trial CME eruption tests!

Might there be peaceful uses for such a capability? In the extremely hypothetical case that a yet greater super-scale CME could be triggered towards a given target in space, such a pulse of denser than naturally possible gas might be captured by a giant braking array designed for such a purpose to provide huge stocks of hydrogen and helium at an asteroid or moon lacking these materials for purposes of future colonization.

A worse weapon on the scale we postulate might be an asymmetric eruption (a form of directed thermonuclear blast using solar hydrogen as thermonuclear fuel), which shoots out a coherent (in the sense of remaining together) burst of plasma at a given target without going runaway and consuming the outer layers of the Sun. If this quite unlikely capability were possible at all (dispersion issues argue against it—but before CMEs were discovered, they too would have seemed unlikely), such an apocalyptic 'demo' would certainly be sufficient emphasis on a threat, or a means of warfare against a colonized solar system. With a sufficient thermonuclear burn –and if the condition of nondispersion is fulfilled—might it be possible to literally strip a planet—Venus, say—of its' atmosphere? (It might require a mass of fusion fuel— and a hugely greater non-fused expelled mass comparable in total to the mass to be stripped away on the target planet.)

It is not beyond the limit of extreme speculation to imagine an expulsion of this order sufficient to strip Jupiter's gas layers off the 'Super-Earth' within. —To strip away 90% or more of Jupiter's mass (which otherwise would take perhaps ~400 Earth years of total solar output to disassemble with perfect efficiency and neglecting waste heat issues). It would probably waste a couple Jupiter masses of material (dispersed hydrogen and helium). It would be an amazing engineering capability for long term space colonization, enabling substantial uses of materials otherwise unobtainable in nearly all scenarios of long term space civilization.

Moving up on the energy scale-- "boosting" or "damping" a star, pushing it into a new metastable state of greater or lesser energy output for times not short compared with the history of civilization, might be a very welcome capability to colonize another star system—and a terrifying reason to have to make the trip.

And of course, in the uncontrollable case of an induced star explosion, in a barren star system it could provide a nebula for massive mining of materials to some future super-civilization. It is worth noting in this connection that the Sun constitutes 99.86 percent of the material in the Solar System, and Jupiter another .1 percent. Literally a thousand Earth masses of solid (iron, carbon) building materials might be possible, as well as thousands of oceans of water to put inside space colonies in some as yet barren star system.

But here in the short-term future, in our home solar system, such a capability would present a terrible threat to the survival of humanity, which could make our own solar system completely barren.

The list of possible countermeasures does not inspire confidence. A way to interfere with the reaction (dampen it once it starts)? It depends on the spread time, but seems most improbable. We cannot even stop nuclear reactions once they take hold on Earth—the time scales are too short.

Is defense of the Sun possible? Unlikely—such a task makes missile defense of the Earth look easy. Once a gravity assist Jupiter flyby nearly stills the velocity with which a flight article orbits the Sun, it will hang relatively motionless in space and then begin the long fall to fiery doom. A rough estimate yields only one or two weeks to intercept it within the orbit of Mercury, and the farther it falls the faster it goes, to science fiction-like velocities sufficient to reach Pluto in under six weeks before it hits.

A perimeter defense around the Sun? The idea seems impractical with near term technology.

The Sun is a hundred times bigger sphere than Earth in every dimension. If we have 10,000 ready to go interceptor satellites with extreme sunshields that function a few solar radii out each one must be able to intercept with 99% probability the brightening light heading toward its' sector of the Sun over a circle the size of Earth, an incoming warhead at around 600 km/sec.

If practical radar range from a small set is considered (4th power decline of echo and return) as 40,000 km then only 66 seconds would be available to plot a firing solution and arm for a destruct attempt. More time would be available by a telescope looking up for brightening, infalling objects—but there are many natural incoming objects such as meteors, comets, etc. A radar might be needed just to confirm the artificial nature of the in-falling object (given the short actuation time and the limitations of rapid storable rocket delta-v some form of directed nuclear charge might be the only feasible countermeasure) and any leader would be reluctant to authorize dozens of nuclear explosions per year automatically (there would be no time to consult with Earth, eight light-minutes away—and eight more back, plus decision time). But the cost of such a system, the reliability required to function endlessly in an area in which there can presumably be no human visits and the price of its' failure, staggers the mind. And such a 'thin' system would be not difficult to defeat by a competent aggressor...

A satellite system near Earth for destroying the rockets moving to the Sun may be a better solution, but with more complications, especially since it would by definition also constitute an effective missile defense and space blockade. Its' very presence may help spark a war. Or if only partially complete but under construction, it may invite preemption, perhaps on the insane scale that we here discuss...

Astronomers see the explosion of stars. They name these stars novae and supernovae—"New Stars" and try to explain (correctly, we are sure, in nearly all cases) their explosion by natural causes. But some few of them, from unlikely spectral classifications, may be result of war between civilizations or fanatic dictators inflicting their final indignity upon those living on planets of the given star. We have enough disturbed people, some in positions of influence in their respective nations and organizations and suicide oriented violent people on Earth. But a nuclear bomb can destroy only one city. A dictator having possibility to destroy the Solar System as well as Earth can blackmail all countries—even those of a future Kardashev scale 2 star-system wide civilization-- and dictate his will/demands on any civilized country and government. It would be the reign of the crazy over the sane.

Author A.A. Bolonkin already warned about this possibility in 2007 (see his interview http://www.pravda.ru/science/planet/space/05-01-2007/208894-sun_detonation-0 [4] Ch.10, Ref. [15] (in Russian) (A translation of this is appended at the end of this article) and called upon scientists and governments to research and develop defenses against this possibility. But some people think the artificial explosion of Sun impossible. This led to this current research to give the conditions where such detonations are indeed possible. That shows that is conceivably possible even at the present time using current rockets and nuclear bombs—and only more so as the centuries pass. Let us take heed, and know the risks we face—or disprove them.

The first information about this work was published in [4] Ch.10, Ref. [15]. This work produced the active Internet discussion in [4] Ch.10, Ref. [19]. Among the raised questions were the following:

1) It is very difficult to deliver a warhead to the Sun. The Earth moves relative to the Sun with a orbital velocity of 30 km/s, and this speed should be cancelled to fall to the Sun. Current rockets do not suffice, and it is necessary to use gravitational maneuvers around planets. For this reason (high delta-V (velocity changes required) for close solar encounters, the planet Mercury is so badly investigated (probes there are expensive to send).

Answer: The Earth has a speed of 29 km/s around the Sun and an escape velocity of only 11 km/s. But Jupiter has an orbital velocity of only 13 km/sec and an escape velocity of 59.2 km/s. Thus, the gravity assist Jupiter can provide is more than the Earth can provide, and the required delta-v at that distance from the Sun far less—enough to entirely cancel the sun-orbiting velocity around the Sun, and let it begin the long plunge to the Solar orb at terminal velocity achieving Sun escape speed 617.6 km/s. Notice that for many space exploration maneuvers, we require a flyby of Jupiter, exactly to achieve such a gravity assist, so simply guarding against direct launches to the Sun from Earth would be futile!

2) Solar radiation will destroy any a probe on approach to the Sun or in the upper layers of its photosphere.

Answer: It is easily shown, the high efficiency AB-reflector can full protection the apparatus. See [1] Chapters 12, 3A, [2] Ch.5, ref. [9]-[12].

3) The hydrogen density in the upper layers of the photosphere of the Sun is insignificant, and it would be much easier to ignite hydrogen at Earth oceans if it in general is possible.

Answer: The hydrogen density is enough known. The Sun has gigantic advantage – that is PLASMA. Plasma of sufficient density reflects or blocks radiation—it has opacity. That means: **no radiation losses in detonation**. It is very important for heating. The AB Criterion in this paper is received for PLASMA. Other planets of Solar system have MOLECULAR atmospheres which passes radiation. No sufficient heating – no detonation! The water has higher density, but water passes the high radiation (for example γ-radiation) and contains a lot of oxygen (89%), which may be bad for the thermonuclear reaction. This problem needs more research.

Summary

This is only an initial investigation. Detailed supercomputer modeling which allows more accuracy would greatly aid prediction of the end results of a thermonuclear explosion on the solar photosphere.

Author invites the attention of scientific society to detailed research of this problem and devising of protection systems if it proves a feasible danger that must be taken seriously.

References

Many works noted below the reader can find on site Cornel University <http://arxiv.org/>, sites <http://www.vixra.org> , <http://www.archive.org> seach "Bolonkin" , <http://bolonkin.narod.ru/p65.htm> and in Conferences 2002-2006 (see, for example, Conferences AIAA, <http://aiaa.org/> , search "Bolonkin") .

1. Bolonkin A.A., (2006)"Non-Rocket Space Launch and Flight", Elsevier, 2005, 408 p.
   http://www.archive.org/details/Non-rocketSpaceLaunchAndFlight,

2. Bolonkin A.A. (2006), Book "*New Concepts, Ideas and Innovation in Aerospace*", NOVA, 2008. .
http://www.archive.org/details/NewConceptsIfeasAndInnovationsInAerospaceTechnologyAndHumanSciences

3. Bolonkin A.A. (2007), "Macro-Engineering: Environment and Technology", pp. 299-334, NOVA, 2008.
   http://Bolonkin.narod.ru/p65.htm, .
   http://www.archive.org/details/Macro-projectsEnvironmentsAndTechnologies

4. Bolonkin A.A. (2008), "New Technologies and Revolutionary Projects", Scribd, 2008, 324 pgs,
   http://www.archive.org/details/NewTechnologiesAndRevolutionaryProjects .

5. Bolonkin A.A., Space Elevator. This manuscript was presented by author as paper IAC-02-V.P.07 at the World Space Congress-2002, Oct.10-19, Houston, TX, USA and published in *JBIS*, vol. 56, No. 7/8, 2003, pp. 231–249. See also Bolonkin A.A., (2005) "Non-Rocket Space Launch and Flight", Elsevier, 2005, Ch.1, http://www.archive.org/details/Non-rocketSpaceLaunchAndFlight,

6. Bolonkin A.A., Man in space without space suite, American Journal of Engineering and Applied Sciences, Vol.2, #4, pp.573-579, 2009. .Published also in [1] Ch.19 (2005); in [4] Ch.6, (2009).

7. Bolonkin A.A., Electrostatic levitation and artificial gravity. Presented as paper AIAA-2005-4465 at 41 Propulsion Conference, 10–13 July 2005, Tucson, Arizona, USA. See also [1], Ch.15 pp.281-302.

8. Bolonkin A.A., A New Method of Atmospheric Reentry for Space Ships. Presented as paper AIAA-2006-6985 to Multidisciplinary Analysis and Optimization Conference, 6-8 Sept. 2006, Portsmouth. See also [2] Ch.8, pp.153-160.

9. Bolonkin A.A., Inflatable Dome for Moon, Mars, Asteroids and Satellites. Presented as paper AIAA-2007-6262 to AIAA Conference "Space-2007", 18-20 September 2007, Long Beach. CA, USA. Published as Inflatable Dome for Moon, Mars, Asteroids and Satellites in http://arxiv.org http://arxiv.org/ftp/arxiv/papers/0707/0707.3990.pdf . See also [3] Ch.2, pp.35-50.

10. Bolonkin A.A., AB Method of Irrigation on planet without Water (Closed-loop Water Cycle). Presented in electronic library of Cornel University http://arxiv.org in 27 December 2007. http://arxiv.org/ftp/arxiv/papers/0712/0712.3935.pdf . See also [3] Ch.1, pp. 3-34.

11. Bolonkin A.A., Artificial Explosion of Sun and AB-Criterion for Solar Detonation. See [4] Ch.10.

12. Bolonkin A.A., (2006). Electrostatic AB-Ramjet Space Propulsion. http://arxiv.org/ftp/physics/papers/0701/0701073.pdf

13. Bolonkin A.A., (2009). Magnetic Space AB-Accelerator. Book "Femtotechnology and Innovative Projects", Lulu, USA. Ch.8.

14. Bolonkin A.A., (2009).Converting of any Matter to Nuclear Energy by AB-Generator and Aerospace. http://www.archive.org/details/ConvertingOfAnyMatterToNuclearEnergyByAb-generatorAndAerospace. American Journal of Engineering and Applied Science, Vol. 2, #4, 2009, pp.683-693. http://www.scipub.org/fulltext/ajeas/ajeas24683-693.pdf .

15. Bolonkin A.A., (2009). Femtotechnology: Design of the Strongest AB-Matter for Aerospace. http://www.archive.org/details/FemtotechnologyDesignOfTheStrongestAb-matterForAerospace . American Journal of Engineering and Applied Science, Vol. 2, #2, 2009, pp.501-514.

16. Bolonkin A.A., (2009). Magnetic Space AB-Accelerator. Book: "Femtotechnology,,," USA, Lulu, Ch.8

17. Bolonkin A.A., Magnetic Suspended AB-Structures and Moveless Space Satellites. http://www.scribd.com/doc/25883886 .

18. Bolonkin A.A., LIFE. SCIENCE. FUTURE (Biography notes, researches and innovations). Scribd, 2010, 208 pgs. 16 Mb. http://www.archive.org/details/Life.Science.Future.biographyNotesResearchesAndInnovations,

19. BolonkinA.A., Robot as Person. Personhood. Three Prerequisites or Laws of Robots. http://www.archive.org/details/RobotAsPersonpersonhood.ThreePrerequisitesOrLawsOfRobots

20. Bolonkin A.A., Krinker M., Magnetic-Space-Launcher. http://arxiv.org/ftp/arxiv/papers/0903/0903.5008.pdf .

21. Pensky O.G., Monograph *Mathematical Models of Emotional Robots*, Perm, 2010, 193 ps (in English and Russian) (http://arxiv.org/ftp/arxiv/papers/1011/1011.1841.pdf).

22. Pensky O.G., K. V. Chernikov*Fundamentals of Mathematical Theory of Emotional Robots* " (http://www.scribd.com/doc/40640088/**).**

23. *Wikipedia.* Some background material in this article is gathered from Wikipedia under the Creative Commons license. http://wikipedia.org .

Projects of space ship and Space Tower

www.ingramcontent.com/pod-product-compliance
Lightning Source LLC
Chambersburg PA
CBHW080915170526
45158CB00008B/2114